Green Chemistry: Theory and Practice

Green Chemistry: Theory and Practice

Paul T. Anastas

Office of Pollution Prevention and Toxics,
US Environmental Protection Agency, Washington, DC, USA

John C. Warner

Department of Chemistry,
University of Massachusetts, Boston, USA

OXFORD
UNIVERSITY PRESS

OXFORD
UNIVERSITY PRESS
Great Clarendon Street, Oxford OX2 6DP
Oxford University Press is a department of the University of Oxford.
It furthers the University's objective of excellence in research, scholarship,
and education by publishing worldwide in
Oxford New York
Athens Auckland Bangkok Bogotá Buenos Aires Calcutta
Cape Town Chennai Dar es Salaam Delhi Florence Hong Kong Istanbul
Karachi Kuala Lumpur Madrid Melbourne Mexico City Mumbai
Nairobi Paris São Paulo Singapore Taipei Tokyo Toronto Warsaw
with associated companies in Berlin Ibadan
Oxford is a registered trade mark of Oxford University Press
in the UK and in certain other countries
Published in the United States
by Oxford University Press Inc., New York

First published 1998
First published new as paperback 2000

A catalogue record for this book is available from the British Library
Library of Congress Cataloging in Publication Data
(Data available)
ISBN-13 978-0-19-850234-0; 978-0-19-850698-0 (pbk.)
ISBN 0-19-850234-5; 0-19-850698-8 (pbk.)
Printed in the United States of America

Preface

Green chemistry is not different from traditional chemistry inasmuch as it embraces the same creativity, and innovation that has always been central to classical chemistry. Green chemistry merely pursues those same ideals with additional considerations to those incorporated into the design and implementation of chemistry historically. These considerations, described in this book, reflect the power that the chemist holds not only over the disposition of the chemistry that is created, but also over all of the implications of the chemistry, from its creation, to its use, until its destruction and beyond. Beyond, because a chemist can not only design a substance to have certain characteristics during its useful life, but can also design what that substance will become (or break down into) after its useful life is over.

This book is not a moral judgment on chemistry but it does elucidate the obligations that chemists, as scientists, have in making choices when designing chemical methodologies. Chemistry itself can be neither 'good' nor 'bad' in a moral sense, as it is merely a natural phenomenon following physical laws. Chemists, however, possess the knowledge and skills to make decisions in the practice of their trade that can result in immense benefit to society or cause harm to life and living systems and they therefore have responsibility for the character of the decisions made. So, while the science of chemistry can be neither holy nor evil, people of either amoral, ignorant, or irresponsible character have misused chemistry and have created a popular disdain for the 'central science' and those who make it their trade.

Basic research in green chemistry is needed. The discovery and development of fundamental chemical transformations that are not harmful to the environment will be the driving force that moves this area forward. Applications of these discoveries will be and have been utilized both for economic and scientific reasons. These methodologies have the potential to affect every aspect of life, just as the field of chemistry has done in the past. Because a synthetic methodology is not sentient, it does not know if it is going to wind up making a pharmaceutical, a paint, or a food additive and thus have a positive impact on all of those chemical products.

It is the chemist who makes these discoveries. It is the chemist who creates the tools, the synthetic methods, that are used throughout industry. Ultimately, because of this role, it is the chemist who has the responsibility for the character of the tools that are in the toolbox. Fortunately for society, it is these same chemists who are solely, uniquely qualified to make those decisions and those discoveries. Green chemistry utilizes the same skills that chemists have always used throughout the history of the science. This book strives to provide a basis and a framework for pursuing the science in the most creative, innovative, and responsible manner possible.

Boston P.T.A.
December 1997 J.C.W.

Contents

1 *Introduction*

1.1 The current status of chemistry and the environment

The status of chemistry in society is a profound dichotomy of perceptions, and neither of these perceptions are in consistent agreement with the facts. While those engaged in the science and industry of chemistry hail the accomplishments that have come from the central science, there are a large number of people who view chemicals and chemistry as something to be afraid of, curtailed, and avoided wherever possible. Neither of these perspectives can possibly capture the full vision of chemistry because it encompasses the characterization, interaction, and manipulation of all matter. The true nature of chemistry, therefore, is complex and vast, as is its effect.

Chemistry has resulted in the medical revolution of the past century in which drugs such as antibiotics have been used to cure diseases that have ravaged mankind for millennia. These advances, brought about by chemistry, have resulted in the average life expectancy rising from 47 in 1900, to 75 years in the 1990s.[1] The world's food supply has seen an explosive expansion in this century because of the development of chemicals that protect crops and enhance growth. In virtually every arena and every aspect of material life – transportation, communication, clothing, shelter, etc. – chemistry has resulted in an improvement, not merely in the trappings of life, but also in the quality of the lives of the billions of individuals who now inhabit the planet.

These almost unbelievable achievements have come at a price. That price is the toll that the manufacture, use, and disposal of synthetic chemicals have taken on human health and the environment. The United States keeps a record of releases of toxic chemicals to the environment by industry, called the Toxics Release Inventory (TRI). The TRI,[2] established under the Emergency Planning and Community Right-to-Know Act, tracks the release of chemicals and classes of chemicals by a variety of sectors of industry on a facility-by-facility basis. While this process provides extremely useful information that was not known prior to its inception in 1986, the chemicals that it covers are only a small fraction of the approximately 75 000 substances in commercial use today and so the amounts of chemicals released to the environment are staggering.

For the 1994 reporting year, there were 2.26 billion pounds of the approximately 300+ hazardous substances released to the environment. At that rate, in the time that it would take you to read one page of this book, over one ton of hazardous waste will have been released to the air, water, and land by industry. The greatest release of hazardous waste to the environment is the chemical industry (Fig. 1.1). Of the top ten industrial sectors whose releases are tracked by the TRI, the chemical industry, including metals, releases more pounds of waste to the environment than the other nine industrial sectors combined.

1.2 Evolution of the environmental movement

1.2.1 Public awareness

It is only fairly recently that the issue of the 'environmental impact' of chemical substances has come into the public dialogue and been fully recognized as a problem. In the years following World War II, there were little or no environmental regulations to speak of that effected the manner in which chemical substances could be manufactured, used, or disposed of. It wasn't until the late 1950s and early 1960s that concern developed over how chemical substances may cause harm to human health and the environment.

In 1962, Rachael Carson wrote the book *Silent Spring*, which detailed the effects of certain pesticides on the eggs of various birds.[3] It illustrated how the use of DDT and other pesticides could spread throughout the food chain, causing irreparable and unanticipated harm. It was the unanticipated nature of the harm that caused a public outcry and resulted in regulatory controls on pesticides which are manufactured and used in the United States.

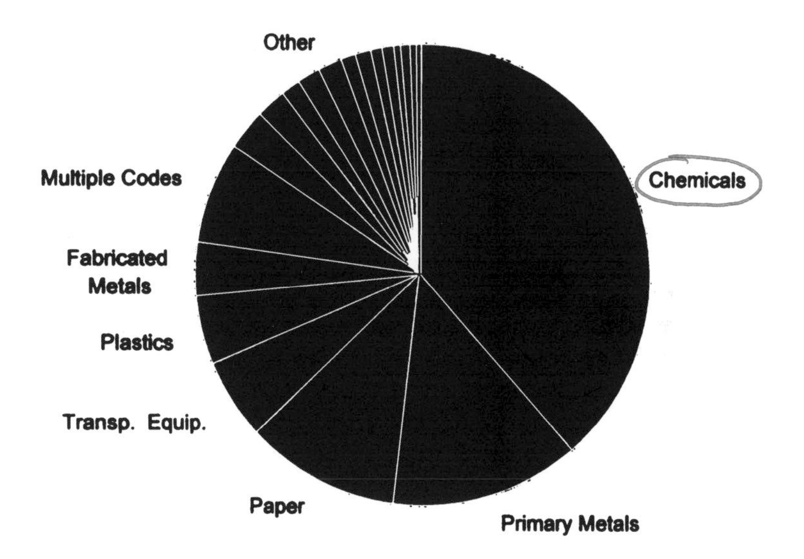

Fig. 1.1. Proportion of hazardous waste released to the environment by the major industrial sectors.[4]

In 1961, there was a scare in Europe about a substance called thalidomide, which was used by pregnant women to lessen the effects of nausea and vomiting during pregnancy ('morning sickness'). As a result of using this drug, the children of the women taking the drug suffered acute birth defects, in many cases in the form of missing or grossly deformed limbs. About 10 000 such children were born world-wide, with 5000 in Germany alone. (Doubts concerning the drug's safety had prevented its sale in the United States.) The tragedy led to stringent governmental regulations for testing new drugs for teratogenic (malformation-inducing) hazards. These 'thalidomide babies', as they are sometimes referred to, caused a great deal of fear in the general public about the effects of synthetic chemicals and the unintended effects that they could have on humans.

In both of the above cases, the public was well aware that the substances in question were designed by scientists, people that the public felt had a great deal more knowledge than they about the chemicals that were being made. Despite the confidence that they had placed in the scientists, to provide innovations for society, the public began to realize that unintended and catastrophic consequences could result from the use of chemical substances. It was unclear to the public whether anyone could control the effects of these substances, and the result was that the government was brought in to control these substances through the regulatory process.

In the subsequent decades of the 1960s, 1970s, and 1980s, a pattern emerged: an environmental problem would manifest itself, where chemical substances were having adverse effects well beyond their intended use, a vocal public outcry would follow, and laws and regulations would be generated to govern and address the problem of chemicals in the environment.

> Times Beach (1983 est. pop. 2000) is a small town in Missouri 40 km (25 mi) south-west of St Louis. Settled in the 1920s as a summer resort for working class St Louis residents, Times Beach soon became a permanent community of small homes and trailer parks. In 1982 the soil along the roads in Times Beach was found to be contaminated with the toxic chemical dioxin. [The town was one of at least 26 and perhaps as many as 100 sites in Missouri that may have been contaminated when dioxin-tainted waste oil was sprayed on the roads a decade ago. The level of dioxin in the soil at these sites varied from around 300 to 740 parts per billion. The federal Center for Disease Control (CDC) rates soil with dioxin readings of over one part per billion as unsafe for long-term contact.] The problems of Times Beach residents were compounded by a flood in late 1982, which forced about 700 families to leave their homes. Government officials urged residents not to attempt to clean up the contaminated mud and debris that had been deposited in their homes by the flood. The federal government provided temporary shelter and, in an unprecedented decision, arranged to buy the entire town, using S33 million from the special fund for toxic waste clean-up.

In addition to the Times Beach situation there were other environmental disaters occurring in the same time-frame. One that caught the public's attention was Love Canal.

> Long-term contamination was involved in the disastrous events at the Love Canal in Niagara Falls, NY. A chemical and plastics company had used an old canal bed in this area as a chemical dump from the 1930s to the early 1950s. The land was given to the city of Niagara Falls in 1953, and a new school and a housing tract were built on it. In 1971, the chemical substances that had been dumped there years before began leaking through the clay cap that sealed the dump, and the area was contaminated by at least 82 chemicals, including a number of suspected carcinogens: benzene, some chlorinated hydrocarbons, and dioxin. Health effects that were linked to the chemical exposure at Love Canal included high birth defect and miscarriage rates, as well as liver cancer and a high incidence of seizure-inducing nervous disease among the neighborhood children. The region was declared an official disaster area. The state paid $10 million to buy some of the homes and another $10 million to try to stop the leakage. About 1000 families had to be relocated. Portions of the site were cleaned up sufficiently by 1990 for houses located there to be put up for sale.

Both the Times Beach and Love Canal events caused sufficient public dismay to prompt the United States Congress to pass new laws to deal with the particular problems that were of the highest visibility. Congress passed the law that became known as Superfund, which would require the clean-up of designated toxic waste sites throughout the country.

Environmental disasters have often resulted in new, and specific, laws being enacted. In the early 1970s, the Cuyahoga River in Ohio was so acutely polluted that it caught fire. The sight of a major river in flames because of chemical pollution prompted calls for legislation to ensure clean water controls through regulation. Nightly news reports showing the brown haze of the Los Angeles or Pittsburgh skylines of the 1960s and 1970s resulted in a variety of Clean Air legislation, including the Clean Air Act. In the 1980s, as the nature of the impact of chlorofluorocarbons (CFCs) on the stratospheric ozone layer became clearer through the use of satellite photographs and the work of Nobel laureate chemists Rowland and Molina, the Montreal Protocol which first called for CFCs to be phased out was adopted. Accidental chemical disasters such as the tragedy at Bhopal, India, where hundreds of people were killed as a result of an accident at a Union Carbide plant that generated the extremely toxic methyl isocyanate, generated the Emergency Planning and Community Right-to-Know Act of 1986.

All of these examples are of unforeseen chemical consequences resulting in tragedy, and the tragedy resulting in public outrage, and the public outrage resulting in legislation to control the manufacture, use, or disposal of chemical substances. But the question should be asked, what has been the nature of these new laws to control chemical substances and are they the only way or the most effective way of protecting human health and the environment from unanticipated outcomes?

1.2.2 'Dilution is the solution to pollution'

During the period prior to the advent of laws that control the release of chemical substances to the environment and significant exposure to people, it was not uncommon for substances to be released directly to the air, water, and land for final disposal. At the time, it was thought that mere decrease of concentration of a substance in a particular medium would be sufficient to mitigate its ultimate impact. This practice and the underlying thinking was sometimes summarized as 'dilution is the solution to pollution'. As absurd as this philosophy is now known to be, it was espoused at a time when factors such as chronic toxicity, bioaccumulation, and even carcinogenicity were not nearly as well understood as they are now.

1.2.3 Waste treatment and abatement through command and control

As toxicity end-points and environmental effects became better known, environmental laws reflected this knowledge by strictly controlling the amounts of a substance that could be released into any particular receiving stream. This first approach to the 'command and control' method of environmental regulation often uses standards or guideline concentrations, e.g. maximum concentration guideline levels (mcgls), to dictate what levels of a particular chemical can be present in the water without adversely affecting humans or the environment.

One of the major shortcomings of this regulatory approach is that it does not usually consider the synergistic effects of other substances present in the water with that of the regulated substance. If a substance that was regulated at a certain concentration resulted in deleterious effects when present in the water with a second, unregulated, substance at a certain concentration level, the public may not be adequately protected from harmful exposures to these substances. The shortcoming of not being able to regulate synergistic effects adequately is widespread throughout current environmental regulations and is not exclusively a problem with regulating concentrations of a substance through command and control.

As environmental regulations progressed, there was an increased emphasis on either treatment of wastes prior to their release, or abatement of the wastes subsequent to their release, in order to mitigate the risks to human health and the environment. Through the use of treatment technologies ranging from neutralization of acids, to scrubbers for air stack emissions, to incineration, environmental statutes required that wastes be transformed into more innocuous forms in order to minimize the impacts of chemical substances.

1.2.4 Pollution prevention

In 1990, the Pollution Prevention Act (PPA), which evolved from the traditional approach of command and control and treatment and abatement, was passed by the United States Congress.[5] The PPA sets a national environmental policy that states that the option of first choice is to prevent the formation of waste at the source. By utilizing a variety of methodologies and techniques, pollution can be prevented, thereby obviating the need for any further treatment or control of chemical substances (Fig. 1.2).

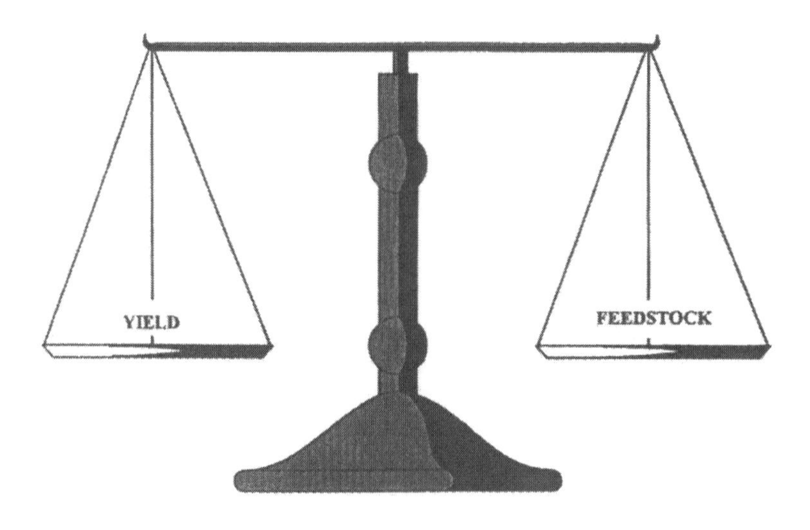

Fig. 1.2. The balance of productivity and resources

Many different ways have been developed to achieve pollution prevention, including engineering controls, which can minimize the amounts of a substance that would otherwise become waste. Inventory controls and techniques, to reduce such problems as unnecessary solvent evaporation, have been extremely successful in reducing the total volume of a chemical substance that ultimately ends up in the waste stream.

1.2.5 Green chemistry

Green chemistry[6,7,8], which is discussed throughout this book, is a particular type of pollution prevention. While there are other methods of achieving pollution prevention that are useful and necessary options, green chemistry is an approach that provides a fundamental methodology for changing the intrinsic nature of a chemical product or process so that it is inherently of less risk to human health and the environment.

Green chemistry involves the design and redesign of chemical syntheses and chemical products to prevent pollution and thereby solve environmental problems. Synthetic chemists play a central role in developing green chemical methods for pollution prevention rather than end-of-pipe control. This textbook focuses on the technical progress and current status of approaches to green chemistry.

Green chemistry is the use of chemical principles and methodologies for source reduction, the most desirable form of pollution prevention. Green chemistry incorporates pollution prevention practices in the manufacture of chemicals and promotes pollution prevention and industrial ecology. Green chemistry is a new way of looking at chemicals and their manufacturing processes to minimize any negative environmental effects. Right now the green chemistry revolution is beginning and it is an exciting time with new challenges for chemists involved with the discovery, manufacture, and use of chemicals.

1.3 The role of chemists

Chemists have always had a role in environmental chemistry. Historically, environmental chemistry meant site monitoring and remediation. Analytical chemists have always had a role in detecting and monitoring environmental problems. Physical chemists have been involved in developing models for environmental phenomena, and atmospheric chemists have studied stratospheric ozone depletion and the greenhouse effect.

Historically, synthetic chemists, those who design new chemicals and their manufacturing processes, have not been particularly environmentally conscious. This is primarily due to the fact that synthetic chemists are at the beginning of the process, identifying ways to make chemicals, while the problems have traditionally been identified with the end of the process, the waste stream. The role of synthetic chemists has historically been to design synthetic pathways to produce target molecules for the least cost in the greatest yield. The public, therefore, has also come to think of synthetic chemists as responsible for the toxic chemicals and for the chemical waste that is generated during chemical manufacturing. It is now known, however, that, while the problems have traditionally been dealt with at the 'end of the pipe', the synthetic chemist upstream can have a major, positive impact on solving these problems.

At present, the costs of a synthetic process for a chemical must include not only those costs for feedstocks and equipment, but also the full cost of regulatory compliance: the cost of waste disposal, liability costs, and treatment costs, including plant modifications for end-of-pipe treatment. These additional costs have driven the total costs of many syntheses to unreasonable levels; full-cost accounting gives new standards for the economics of manufacturing. Chemists have the power to reduce these indirect costs significantly by redesigning chemicals and their processes.

As a chemist puts pencil to paper to design a new chemical synthesis, he or she is making intrinsic decisions: decisions about whether hazardous substances will be used, whether hazardous materials will have to be handled by workers, whether hazardous wastes or by-products will require special disposal, and the like. All of these decisions are inherent in the synthetic process. The goal of green chemistry, or benign chemistry, is to design synthetic methodologies that reduce or eliminate the use or generation of toxic feedstocks, by-products, solvents, and all other associated products. A synthetic chemist who develops a 'green chemistry' synthesis is likely to produce a more cost-effective product when all direct and indirect costs are accounted for.

2 *What is green chemistry?*

2.1 Definition

Green chemistry, environmentally benign chemical synthesis, alternative synthetic pathways for pollution prevention, benign by design: these phrases all essentially describe the same concept.

Green chemistry is the utilization of a set of principles that reduces or eliminates the use or generation of hazardous substances in the design, manufacture and application of chemical products.

Green chemistry is not complicated although it is often elegant. It holds as its goal nothing less then perfection, while recognizing that all of the advances and innovations towards this goal will contain some discrete risk. It is through these continual incremental improvements that the objectives of green chemistry will be realized. Chemists have always striven for efficiency in their synthetic methodologies. To be able to conduct a transformation or construct a pathway to a molecule effectively and reliably was fundamental to the art of the synthetic chemist. Efficiency is important, not only as a measure of the quality of a synthetic method, but also as a practical and economic consideration as well. Economic considerations have played a major role in designing syntheses that use the most available and/or lowest cost feedstock.

Recent additions and enhancements of the above criteria have only added to the definition of synthetic elegance. Stereochemical specificity, for example, serves to augment the insistence on efficiency. The concept of 'atom economy' has replaced the time-honored metric of 'yield' as a standard by which to measure the quality of a synthetic methodology.[9]

Green chemistry recognizes that these standards need to be taken one step further to be complete. Much like the Hippocratic procedures and protocols, a synthetic chemical methodology, to be truly elegant, must 'first, do no harm'. This criterion means that it is incumbent upon synthetic chemists to incorporate the provisions of ultimate effect on human health and the environment into the design of any new methodology.

While it is all too easy to dismiss this criterion as someone else's domain, chemists no longer have this luxury. In the days of taste testing and pipetting by mouth, there may have been a reasonable belief that the level of knowledge of chemical hazards among chemists was too low to warrant responsibility. Those days have long passed. Hazard data being both generated and required by law, the chemists of today are aware of the potential and real effects of their work. It is no more excusable for a fireman not to know that a fire burns, or a chef not to know that a knife cuts, than for chemists not to know the character of the tools of their trade.

Equipped, or burdened, with this knowledge, the chemist must confront responsibilities. No longer can the consequences of the trade be blamed on misuse by businessmen and industrialists. Because chemists possess the understanding of molecular manipulation and have the information necessary to assess how those manipulations may or may not put human health and the environment at risk, they have entered an era where this knowledge must play a central role in the conduct of the trade.

This realization by the purveyors of green chemistry, is being viewed as an opportunity rather than a limitation.

2.2 Why is this new area of chemistry getting so much attention?

To answer this question we must first have a basic understanding of the problem. There has been much debate over the past generation over the exact nature of the environmental hazards that have been generated as a result of the release of various synthetic chemicals into the environment. There is little doubt that until the uncertainties in the toxicological data – exposure, fate, and transport data – and the risk analyses are unequivocally resolved, this debate will continue, probably for at least the next generation. Therefore, there are two logical choices left to the scientific community. The first, is to allow the uncertainties described above to continue to be paralysing and not to attempt to address the concerns for human health and the environment. The second option, which those pursuing the new area of green chemistry have adopted, is to accept the fact that the release of chemicals to the environment causes some incremental increase in the risk to human health and the environment. The risk can be eliminated through fundamental breakthroughs in chemical methodologies that are technically and economically viable, then the chemical community should pursue it.

It is as true in green chemistry as it is in every other area of science that one can only operate with the current state of knowledge, but, with the knowledge of chemical hazards that now exists, chemists can aim to minimize them.

2.3 Why should chemists pursue the goals of green chemistry?

Robert F. Kennedy has been quoted as saying 'Some see things as they are, and ask why. I see things as they should be and ask why not'. One of the most basic philosophical reasons that chemists must try to make the work they do and the substances they use as environmentally benign as possible is that we can. With knowledge of how to manipulate and transform chemicals, coupled with the basic hazard data that can be accessed readily from a variety of sources, chemists have it in their power to reduce or eliminate the risk posed to themselves and society in general by the chemical enterprise.

Researchers at the vanguard of innovation in this new area know that these goals can be accomplished. While everyone understands that no activity can be completely risk free, the goals achieved both at the research bench and in commercialized processes have greatly decreased environmental and health concerns, while developing efficacious process and methodologies.

Yet another reason for the chemistry community to pursue green chemistry vigorously is because it is based on fundamental molecular science providing the root of the solution, rather than on applying a bandage or patchwork approach to risk reduction. At its most basic level, risk can be described by the following formula in Fig. 2.1.

Risk = f[hazard,exposure]

Fig. 2.1. Formula describing risk.

The traditional way that industry and society, through national environmental policy, has dealt with the reduction of risk is through the reduction of exposure. With a fixed hazard and a reduced exposure, risk should decrease proportionately. By using well-characterized hazards (toxicity data) and knowing the efficacy of whatever exposure control method is being used, the risk can be manipulated until it is below some identifiable, acceptable level. This 'acceptable level', by necessity, has to be arbitrary since the question of 'acceptable to whom?' must then be faced.

While a 1:1 000 000 cancer risk from exposure to a certain level of a substance may be defined by society in its laws and regulations as 'acceptable', it certainly is not acceptable if you are that '1' in 1:1 000 000.

Another limitation of exposure controls to reduce risk is that the use and release of a chemical may affect individuals who aren't using those controls. For instance, a chemical worker may be wearing gloves, goggles, etc. in order to protect themselves from being exposed to high levels of a certain substance known to have acute effects. But how do these exposure controls affect someone downstream or downwind who is not protected by exposure controls. With uncertainties with chronic effects, bioaccumulative effects, and synergistic effects being extremely high for a broad range of substances under the current status of the science, the use of exposure controls to reduce risk to society in general is called into question.

A final reason why exposure controls may be limited is because they can fail. No respirator, face shield, glasses, gloves, goggles, or protective suit is perfect. With failure of this exposure-limiting equipment, the individual is then at maximum risk from the hazard.

Contrast the above limitations to the hazard-reducing principles of green chemistry. The largest difference between the two approaches to risk reduction is that hazard reduction through green chemistry, when done properly, cannot fail. If, through the variety of techniques and methodologies that will be discussed in this book (e.g. alternative feedstocks, solvents) the hazard is reduced, there is no way that the risk can increase through a spontaneous increase in the hazard. In other words, there is no way that an innocuous substance is going to become arbitrarily toxic to human health and the environment. Now, of course, in the same way that someone could put safety goggles on backwards or put protective rubber gloves on their feet, it is equally conceivable that green chemistry can be performed incorrectly. This would be equivalent to substituting an extremely toxic substance for one that is virtually non-toxic. Beyond this absurd exception, hazard reduction through green chemistry must necessarily reduce risk. Also, in contrast to the use of exposure controls to reduce risk, the effect of exposure on downstream or downwind recipients of the chemical hazard is less. Because the intrinsic nature of the substance itself is less hazardous, there is no differential risk to the person working with the substance versus the secondarily exposed individual. Finally, the concept of a level of acceptable risk is eliminated as a target and replaced with the optimal goal of environmentally benign. While more will be said on this topic later, the goal of making a chemical product or process 'environmentally benign' is a mere statement of the ethic of continuous improvement more than it is a metric by which to measure improvement.

It is because of the reasons outlined above that the approach to risk reduction through utilization of hazard reduction via green chemistry is preferable to that via exposure control.

From an economic standpoint, one fact is intuitively obvious. There is not a chemical product or process that is going to be more economically favorable because of the need for, or use of, exposure controls. Regardless of the methods or equipment used, from engineering controls to personal protective gear, there is always an associated cost and that means an economic drain. More will be said on this subject when the full costs of risk reduction are addressed later in this chapter.

In contrast, a green chemistry solution to risk reduction could, potentially, have a variety of economic benefits associated with its implementation. Some possibilities include lower feedstock costs, higher conversion rates, shorter reaction time, greater selectivity, enhanced separations, or lower energy requirements. While not every green chemistry solution is going to have these benefits, these potential economic gains stand juxtaposed to the certain economic drains of exposure controls. It should be noted that the advantages of green chemistry listed relate to direct operating costs. Additional indirect cost advantages will also be discussed later.

2.4 The root of innovation

It has been stated that green chemistry is a 'fundamental' approach to environmental problems caused by pollution. What exactly does that mean? There have been many approaches to dealing with environmental problems around the world. In the modern industrial age, these approaches have evolved as the level of understanding of the problems themselves has evolved. While today, these approaches to pollution may seem as backward as blood-letting does to a physician, it is important to review them briefly in order to see how the dominant paradigms in environmental protection have come into being.

Throughout the years immediately following the end of World War II, industry in the United States and the other industrial nations grew at unprecedented rates. With this development came a significantly higher standard of living, many modern conveniences, and a great deal of pollution. Since the type of tracking of chemical releases to the environment that we have today did not exist then, we can only extrapolate data into the past and correlate it to production volumes to conclude that releases were extremely high. What is, however, most compelling about retrospective analysis of industry is the way in which chemical releases were dealt with.

2.5 Limitations/obstacles

Green chemistry is an approach dealing with one of the fundamental environmental problems of the world, the problem of pollution.

Pollution and its recognition as the basis for many problems, such as destruction of ecosystems and threats to wildlife, has been at the heart of environmental policy since the 1960s. Most of the major pieces of environmental legislation that have been enacted over the past 35 years, such as the Clean Water Act, the Clean Air Act, the Resource Recovery and Conservation Act, the Safe Drinking Water Act, and Superfund in the US, have centered around the problem of pollution.

Pollution, in virtually all of the above-mentioned legal contexts is defined as chemical pollution. The release to the air, water, and land of substances that may cause risk to human health and the environment is the pollution that the environmental policies of the United States have sought to remedy. Most often the approach utilized in the laws has been one of controlling the amount of release of a chemical or controlling the final concentration of that chemical in the medium to which it was released. This approach, while yielding dramatic improvements to the overall quality of the air, water, and land, has met with significant philosophical and logistical obstacles. The obstacles can be best described by the questions posed in Fig. 2.2.

Question: *What is a drum of hazardous waste with a drop of champagne called?*

Answer: *Hazardous waste.*

Question: *What is a drum of champagne with a drop of hazardous waste called?*

Answer: *Hazardous waste.*

At some point when one is dealing with substances of high toxicity, unknown toxicity, carcinogenicity, or chronic toxicity, it becomes problematic, if not impossible, to set appropriate levels that are tolerable to human health and the environment. Green chemistry, however, deals with the above situation somewhat differently.

Over the course of the past generation or two, there has been a large volume of research conducted on elucidating the hazards associated with various chemical substances. Part of the impetus for this research was the need to have data on the toxicity of substances so that levels of acceptable exposure could be determined. When much of the environmental legislation was initially enacted, relatively little toxicological data existed on chemical substances. With the subsequent data generation that has taken place, the approach of green chemistry becomes possible.

Green chemistry utilizes the data available on a large number of compounds in its evaluation of chemical products, chemical syntheses, and the entire life cycle of the chemical enterprise. Through recognition that both identification of which substances are likely to contain a greater hazard for human health and the environment, and increased understanding of the mechanism of action by which this hazard is derived, chemists can consciously design safer chemistry and safer chemicals.

Rather than using the historical approach of controlling the concentrations or releases of a particular chemical substance through engineering controls or waste treatment procedures, green chemistry, instead, changes the intrinsic nature of the substances themselves so as to reduce or eliminate the hazard posed by the substances. This is particularly appealing from a science – philosophy perspective because of the nexus between classical chemistry and mechanistic toxicology. At this nexus, the green chemist can balance the molecular features that are needed for the tasks to be performed with those molecular features that are responsible for the intrinsic hazard of the substance. This balance can be applied equally to the design of an individual target molecule or to an entire synthetic pathway.

As Parcelus said, many years ago, 'Everything is a poison, depending on the dose'. It would be easy to use this as a justification for just throwing up one's hands and saying well if everything is toxic then there's nothing that a chemist can do. This rationale, while baseless, is often used in order to justify inaction. Merely because everything has some discreet hazard, which is the essence of what Parcelus was saying, that does not mean that everything is equally hazardous. As a matter of fact, levels of toxicity vary over many orders of magnitude. In addition, chemicals have many different toxic end-points and mechanisms of action.

It is a fact that while toxicological data are available on thousands of chemical compounds this represents only a small percentage of the millions of compounds known today. This, however, ignores that fact that correlations within chemical classes have been extensively elucidated, and, while there may not be comprehensive toxicological data on a specific chemical substance, it is both a reasonable and effective first assumption that it will behave similarly to other compounds in the same chemical class.

How does a synthetic chemist become aware of this toxicological data on all of these compounds without becoming a toxicologist and devoting extensive time and energy to finding hazard information rather than on chemical research? There are many readily available data sources on the toxicity of chemicals, ranging from single-volume references to on-line databases. Another approach for utilizing the information as a tool is to look at the lists of chemicals that have been developed by various health, medical, or regulatory agencies. There are also dedicated lists that identify which chemical substances are of highest hazard to human health and the environment.

3 *Tools of green chemistry*

As the environmental and health effects of a chemical or chemical process have begun to be considered at the design stages, the approaches and the development of techniques have been quite diverse. Since the types of chemicals and the types of transformations are so varied, so too are the green chemistry solutions that have been proposed. These, however, can be broken down into several categories.

With alternative synthetic design, we look not at the ultimate molecule but at the synthetic pathway used to create it. By modifying the synthesis, we can arrive at the same final product, yet reduce or eliminate toxic starting materials, by-products, and wastes. Two of the main components of chemical synthesis are feedstocks and reaction conditions: either or both of these may be changed to produce alternative (and improved) chemical syntheses.

3.1 Alternative feedstocks/starting materials

Much of the character of a reaction type or synthetic pathway is determined by the initial selection of the starting materials. Once that selection is made, many options then fall into place as a necessary consequence of that decision. The selection of a feedstock has a major effect, not only on the efficacy of the synthetic pathway, but also on the environmental and health effects of the process. The selection of a feedstock for the manufacture of a product determines what hazards will be faced when workers are handling this substance, suppliers are manufacturing it, and shippers are transporting the substance. With significantly large commodity chemicals, the selection of a particular feedstock may drive the market such that its role as a feedstock is its primary reason for existence. Therefore the selection of the feedstock is a very consequential part of the green chemistry decision-making process.

Embedded in this decision are some very pertinent questions. Do you want to make the product from virgin material or recycled material? Do you want to make the substance from petroleum feedstocks, biological feedstocks, or other alternative feedstocks?

Currently, 98% of all organic chemicals synthesized in the United States are made from petroleum feedstocks. Petroleum refining takes up 15% of the total energy used in the US, and its energy usage is rising because the low quality raw petroleum available now requires more energy for refinement. During conversion to useful organic chemicals, petroleum undergoes oxidation, the addition of oxygen or an equivalent; this oxidation step has historically been one of the most environmentally polluting steps in all of chemical synthesis. As a result of these considerations, it is important to reduce our use of petroleum-based products by using alternative feedstocks.

In general, agricultural and biological feedstocks can be excellent alternative feedstocks. Many are already highly oxygenated and so their use in place of petroleum feedstocks eliminates the need for the polluting oxygenation step (Figs 3.1 and 3.2). Furthermore, syntheses can be accomplished that are significantly less hazardous than when conducted with petroleum products. At present, research has shown that a host of agricultural products can be transformed into consumer products: products such as corn, potatoes, soy, and molasses are being transformed through a variety of processes into such products as textiles, nylon, etc.

The exploration of biological sources of alternative feedstocks need not be limited to agricultural products: agricultural waste or biomass, and non-food-related bioproducts, which are often made up of a variety of lignocellulosic materials, may provide important alternative feedstocks.

Other classes of alternative feedstocks are also emerging, such as light. For example, heavy metals, which are often used in petroleum oxidation processes, are also quite toxic and are carcinogens or cause damage to neurological systems. Recently discovered alternative syntheses replace the heavy metal reagents with the use of visible light to carry out the required chemical transformations.

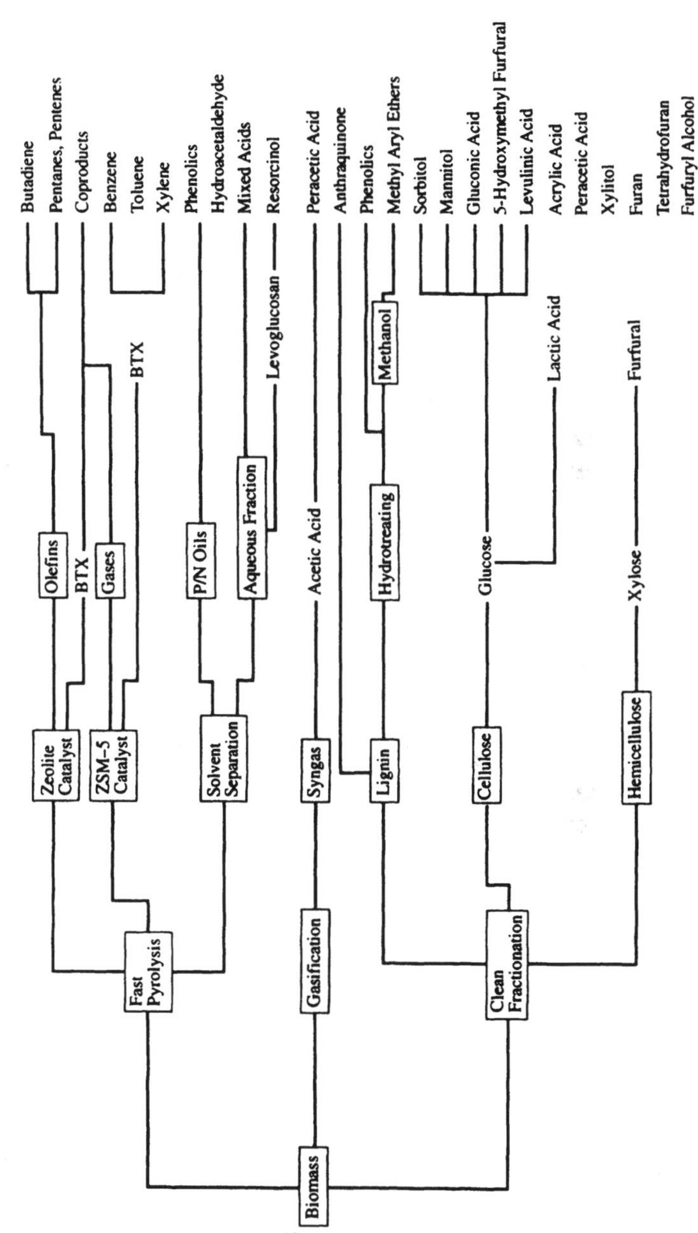

Fig. 3.1. Biological feedstocks.[10]

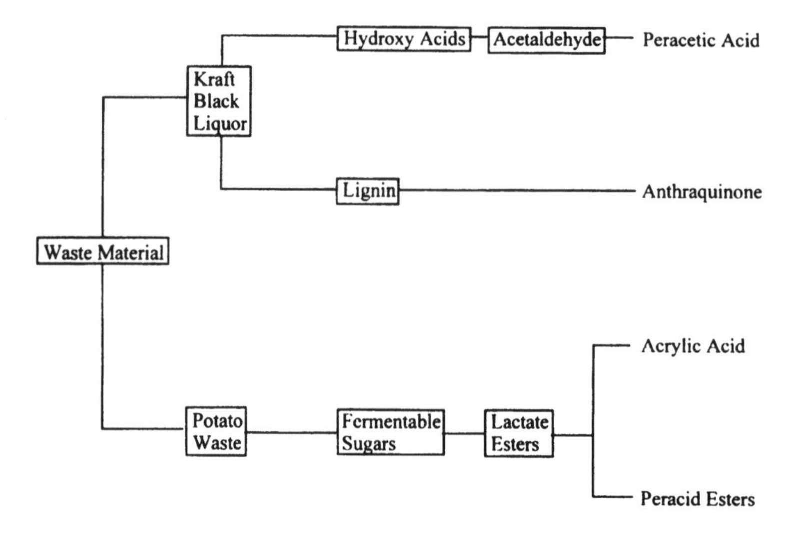

Fig. 3.2. Agricultural feedstocks.[10]

3.2 Alternative reagents

In transforming the selected feedstock into the target molecule, the synthetic chemist has already worked out the structural modifications that are necessary. While the goal of each synthetic step is clear, the reagent needed in order to carry out that step may have yet to be selected. It is at this point that the synthetic chemist must balance the criteria of efficiency, availability, and effect in order to assess the best reagent for carrying out the transformation.

The transformations have to be evaluated to determine whether they are stoichiometric or catalytic, atom economical or not, and what the characteristics of any wastes that will be generated through the use of the reagent will be. The selection of a particular reagent versus another for the same net transformation can have an effect on all of the above factors.

3.3 Alternative solvents

An important area of green chemistry investigations has centered around the selection of a medium in which to carry out a synthetic transformation. Because the dominant paradigm of chemical synthesis has been based around solution chemistry, the question is often phrased as 'What solvent should be used?'. This phrasing, of course, begs the question, 'Should a solvent be used at all?'. Many of the solvents commonly used are some of the volatile organic compounds known to cause smog when released to air. These solvents are listed in the United States' Clean Air Act as substances to be avoided. Research is being conducted that pursues chemistry that has previously been done in a solvent and discovers a way to do the same chemistry in various solventless systems.

Once it has been determined that a solvent is needed or, for whatever reason, is preferable for a particular synthesis, the chemist must select from a number of alternatives. While traditional organic solvents are well known, characterized, and utilized, alternatives such as aqueous systems, ionic liquids, immobilized solvents, dendrimers and amphiphilic star polymers, and supercritical fluids (all of which will be discussed in later chapters) are being increasingly applied in synthesis. Supercritical fluids are gases at a combination of high pressure and low temperature. When environmentally friendly gases are made into supercritical fluids and used as solvents, environmental benefits are obtained. A good example is that of carbon dioxide which is currently being used as a supercritical fluid to provide solvation for a variety of chemical reactions.

While these decisions regarding solvents will ultimately be made on a case-by-case basis by the chemist, the information currently being generated showing the usefulness of these alternatives for a wide variety of chemical classes and reaction types is proving useful to the research chemist and the process development chemist.

3.4 Alternative product/target molecule

While a synthesis is often driven by the pursuit of a particular target molecule, it is also commonly the case that what is actually being pursued is the ability to make any chemical that can serve a particular function or have a certain performance criterion. For many years the pharmaceutical industry has been doing research into designing safer chemicals. With pharmaceuticals, the object is to maximize the therapeutic benefits of a molecule while minimizing or eliminating the toxic side-effects. These same principles can be applied to the full range of chemical applications.

In the cases where function is the primary motivation, molecular manipulation that preserves efficacy of function while mitigating toxicity or other hazards is the goal of green chemistry. Through these efforts and other toxicological research, it is often possible to identify the part or parts of a molecule that produce toxic effects. Similarly, through chemical research, we are able to identify those parts of a molecule that are required to give the chemical the desired function – to allow it to serve a specific, desired use.

In designing safer chemicals, one identifies the undesirable, toxic portion of a molecule and lessens or eliminates its toxicity, while maintaining the function of the molecule. In many cases, the overlap of the toxic and functional portions creates a worthy challenge for the synthetic chemist.

3.5 Process analytical chemistry

By process analytical chemistry we mean real-time measurement of reaction conditions during chemical synthesis, coupled with the ability to alter the reaction depending on the outcome of the analyses. For example, imagine that pollutant X is being generated in a reaction in trace amounts but is formed in larger amounts if the temperature and pressure become too high. Using process analytical chemistry, one could measure the concentration of pollutant X constantly during a reaction and immediately change the reaction conditions if the amount of X becomes unacceptably high. A considerable amount of research is currently being done in the field of process analytical chemistry. The technique is particularly applicable to biotechnological syntheses; the reactions involved are usually quite complex and the value of the resulting product is high, thus making process analytical chemistry cost-effective.

The context in which one evaluates a catalyst can largely determine the outcome of that evaluation. For example, if you evaluate a catalyst when the alternative is a stoichiometric reaction you may get one answer. If you evaluate the same catalyst to assess its toxicity and environmental hazard, you may arrive at a very different answer. Like many areas of pollution prevention and environmental protection, the trade-offs between performance and risk cannot be ignored.

3.6 Alternative catalysts

Some of the major advances in chemistry, especially industrial chemistry, over the past generation have been in the area of catalysis. Catalysis has not only advanced the level of efficiency but has also brought about concurrent environmental benefits. Through the use of new catalysts, chemists have found ways of removing the need for large quantities of reagents that would otherwise have been needed to carry out the transformations, and would ultimately have contributed to the waste stream. It is also true that various classes of catalysts, such as the heavy metal-based catalysts, have been found to be extremely toxic.

4 *Principles of green chemistry*

Green chemistry has been defined concisely in Chapter 2, and while green chemistry has also been described as the Hippocratic Oath for the Chemist ('First, do no harm.'), the true definition of a subdiscipline or an area of investigation comes from the research and the accomplishments that are conducted therein. It is this organically grown definition that not only answers the question 'What is green chemistry?', but also provides the scope and range for green chemistry so that we can view where green chemistry can and will go in the future.

The listing of the 'Twelve Principles of Green Chemistry' should be viewed as a reflection of the science that has been done within this nascent field in the recent past, as well as as a direction that has been set by some of the pioneering scientists who have laid the groundwork for the future (Fig. 4.1).

4.1 It is better to prevent waste than to treat or clean up waste after it is formed

It has always been anticipated and expected that there are 'normal' costs associated with the manufacture and use of chemicals and chemical products. Certainly, you need to pay for the starting materials and reagents that will be an intrinsic part of the product. However, a cost that has become of significant importance over the past 20 years is the cost of treatment and disposal of chemical substances. The more hazardous the substance, the more costly the substance is to deal with. This is generally true whether one is discussing a large chemical manufacturer or a small academic laboratory.

Fig. 4.1. The twelve principles of green chemistry.

1. It is better to prevent waste than to treat or clean up waste after it is formed.
2. Synthetic methods should be designed to maximize the incorporation of all materials used in the process into the final product.
3. Wherever practicable, synthetic methodologies should be designed to use and generate substances that possess little or no toxicity to human health and the environment.
4. Chemical products should be designed to preserve efficacy of function while reducing toxicity.
5. The use of auxiliary substances (e.g. solvents, separation agents, etc.) should be made unnecessary wherever possible and, innocuous when used.
6. Energy requirements should be recognized for their environmental and economic impacts and should be minimized. Synthetic methods should be conducted at ambient temperature and pressure.
7. A raw material of feedstock should be renewable rather than depleting wherever technically and economically practicable.
8. Unnecessary derivatization (blocking group, protection /deprotection, temporary modification of physical/chemical processes) should be avoided whenever possible.
9. Catalytic reagents (as selective as possible) are superior to stoichiometric reagents.
10. Chemical products should be designed so that at the end of their function they do not persist in the environment and break down into innocuous degradation products.
11. Analytical methodologies need to be further developed to allow for real-time, in-process monitoring and control prior to the formation of hazardous substances.
12. Substances and the form of a substance used in a chemical process should be chosen so as to minimize the potential for chemical accidents, including releases, explosions, and fires.

It is staggering to imagine that in many of the large chemical companies in the United States, expenditures on research and development are equal to expenditures on environmental health and safety. In this statement lies the illustration that the one of the true victims of the costs of using and generating hazardous substances is the further growth and innovation of the science and industry of chemistry. Universities and small colleges are meeting the challenge of the cost of waste disposal from chemistry labs, both educational and research, by reducing either the number of laboratories or reducing the scale upon which laboratory experiments are run.

The costs of dealing with hazardous substances, either through handling, treatment, or disposal, have continued to increase substantially. These costs will now have to be factored in unless they are prevented. The only way to prevent these costs arising is to avoid the use or generation of hazardous substances by designing chemistry through the use of green chemistry techniques. It is through this method that the costs of everything, from engineering controls, to personal protective gear, to regulatory compliance, can be minimized, if not avoided, and the associated expenditures prevented.

One type of waste product that is both common and often the most avoidable is starting material or reagent that is unconverted. One tool that allows one to assess the illogicality of wasting starting material under today's cost structures is the simple thought that 'When one wastes starting material, one is paying for the substance twice; once as a feedstock and again as a waste and so one has achieved no utility from the substance'. Often the cost of waste disposal may be many times the cost of the virgin starting material. Put in these terms, the infrugality of the equation is intolerable wherever it is avoidable.

In virtually every aspect in society, it has long been acknowledged that preventing a problem is superior to trying to solve the problem once it has been created. Proverbs going back hundreds of years have said it: 'An ounce of prevention is worth a pound of cure'. A similar analogy is that of preventative medicine, which is widely recognized as being preferable to having to cure the disease.

In contrast, the chemical industry and other manufacturers or processors of chemicals have eschewed prevention for generations until recently. Traditionally, and still in some sectors, the rationale has been that, although certain substances and wastes were hazardous, chemists knew how to handle and treat these chemicals. This rationale is as illogical as saying that because a doctor knows how to cure an affliction, a person should not try to avoid contracting it. The damage and the costs of fixing a problem are always greater than those of preventing it, whether it be appendicitis or toxic chemicals. There will always be hazards and dangers that are unavoidable in life and it is therefore foolhardy to waste time, money, and effort on dealing with those challenges that are avoidable.

A criterion that is often assessed to calculate the need or desire for treatment or control of a chemical substance is whether or not damage has been done by the generation of waste. In many cases the 'damage' is merely that substances have been uselessly processed or transformed, costing energy, money, time for separation from product, and almost always requiring technology to get rid of them or to render them innocuous. In the equation of what harm has been done by the generation of one waste or another, the evaluation often centers simply on human life and health or the well-being of the environment. While these criteria are of paramount concern, it should also be recognized that when avoidable wastes are generated, or in some cases merely where avoidable hazards are used, there is 'harm' and there is 'damage'.

The mere fact that a process generates waste means that separation, treatment, and disposal of the hazardous substances is needed. The fact that hazardous substances are used means that there has to be special handling, protective gear, and engineering controls. In the broadest terms, this is damage, which needs to be factored into the evaluation of what substances and what processes to use. It is possible that in certain cases this 'damage' may not be of enough consequence to change the nature of the process, but it is always of enough consequence to be considered.

4.2 Synthetic methods should be designed to maximize the incorporation of all materials used in the process into the final product

Throughout the 20th century texts on organic chemistry have not featured balanced chemical equations. The reactions they depict seldom, if ever, account for the by-products and coproducts that are necessarily generated in the course of a synthetic transformation. The classic evaluation of the effectiveness and efficiency of a synthesis is yield. 'Yield' also totally ignores the use or generation of any undesirable products that are an intrinsic part of the synthesis. It is possible, and very often the case, that a synthetic pathway, or even a synthetic step, can achieve 100% and generate waste that is far greater in mass and volume than that of the desired product. This is true because the calculation is based around the mole concept of moles of starting material vs. moles of product. If a mole of starting material produces a mole of desired product, the yield is 100% and the synthesis is deemed perfectly efficient by this calculation. This same transformation, however, could produce one or more moles of waste for every mole of product. Each mole of that waste could be many times the molecular weight of the desired product. Therefore a 'perfectly efficient' synthesis according to the percentage yield calculation could generate significant amounts of waste and this would be invisible using only this evaluative equation.

A now classic illustration of this shortcoming in the yield approach is that of the Wittig reaction. The Wittig introduces a functionality with a molecular weight significantly less than the molecular weight of the waste. This reaction may still proceed with a 100% yield. It is because of this discrepancy that the concept of atom economy is used. Atom economy is an assessment in which one looks at all of the reactants to measure the degree to which each of them is incorporated into the final product. Therefore, if all the reactants are incorporated into the product completely, the synthetic pathway is said to be 100% atom economical. The standard synthetic transformation types can be evaluated generically to determine the intrinsic atom economy of each type.

4.2.1 Rearrangements

By definition, a rearrangement reaction is a reorganization of the atoms that make up a molecule. Therefore, by necessity, it is a 100% atom economical reaction, where all the reactants are incorporated into the product.

4.2.2 Addition

Because addition reactions add the elements of the reactant to a substrate with total inclusion (e.g. cycloadditions, bromination of olefins) they are atomic economical.

4.2.3 Substitution

When a substitution reaction is effected, the substituting group displaces a leaving group. The leaving group is necessarily a waste product of the reaction that is not included in the final product and therefore diminishes the atom economy of the transformation. The exact degree to which the reaction is non-atom economical is dependent on the specific reagents and substrates used.

4.2.4 Elimination

Elimination reactions transform the substrate by reducing the atoms to generate the final product. In this case, any reagents used do not become part of the final product and the eliminated atoms are lost as waste. This is, therefore, intrinsically the least atom economical of the basic synthetic transformations.

4.3 Wherever practicable, synthetic methodologies should be designed to use and generate substances that possess little or no toxicity to human health and the environment

The fundamental basis of green chemistry is the incorporation of hazard minimization or elimination into all aspects of the design of the chemistry.

Unlike so many attempts in the past to protect the environment through the limitation, regulatation, or elimination of chemistry and chemicals, however, the green chemistry approach actually embraces chemistry as the solution rather than the problem. Green chemistry recognizes that it is through the skills and knowledge possessed by chemists that today's world can have the modern technological advances that have come to be expected from the scientific community in a way that is safe for human health and the environment.

There is an intrinsic need to consider hazard when designing chemistry for the environment. There are only two ways to minimize risk of harm of any kind: either minimize the exposure or minimize the hazard. Minimizing exposure can take a variety of forms, such as protective clothing, engineering controls, respirators, etc. Those who feel that hazard should not be included in an evaluation and design of chemistry often believe that since 'they know best how to handle hazards' because of their chemical expertise, they should use whatever substance they choose regardless of the hazard (e.g. toxicity, flammability).

The pragmatic reason that hazard must be considered is twofold. First, it is virtually impossible to control exposure without increasing the cost of a process. All of the control mechanisms, whether they be clothing, engineering controls, etc., cost something. One is therefore adding cost unnecessarily. Secondly, exposure controls can fail and so ones risk increases commensurately with that failure. Hazard, however, is an intrinsic characteristic that isn't going to change and therefore the risk will not increase spontaneously.

There are additional answers to the question, 'Why should we consider hazard?' From an ethical standpoint, one answer is, 'because we can'. Chemists have the knowledge and skills to minimize the hazard faced by the public, the environment, and users of chemicals in general. With possession of that knowledge, there is a responsibility to ensure that no harm is done wherever practicable. That does not mean eschewing knowledge to avoid all possible harm, it means using the knowledge chemists already have to pursue future innovations in ways that are safest for human health and the environment.

An equally valid answer to the question is, 'because we must'. We must, because, whether you look at it from an environmental, economic, legislative, or social perspective, chemistry, chemists, and the chemical industry have no choice. The environment has sustained tremendous damage because of misuse and the lack of vision of some members of the chemical enterprise and society in general. The chemical industry and universities are being bled dry economically by the costs of dealing with these hazards. Legislatively, increasingly restrictive laws have the potential to strangle innovation if scientific solutions do not displace them. And, socially, the role of chemists has become a dichotomy of being considered both innovators and polluters. None of these circumstances, which generally describe the status quo, are acceptable. By designing chemistry that reduces or eliminates hazard, green chemistry offers the scientific option to deal with each of those problematical circumstances.

4.4 Chemical products should be designed to preserve efficacy of function while reducing toxicity

4.4.1 What is designing safer chemicals?

This area of green chemistry is often referred to simply as 'designing safer chemicals'. Through knowledge of the molecular structure, chemists can obviously determine quite a bit about the characteristics of the compound. Certainly, the field of chemistry has developed extensive tools by which to measure and estimate the properties of chemical products. These products can be as diverse as dyes, paints, adhesives, or pharmaceuticals, and the properties assessed as varied as color, tensile strength, cross-linking potential, and antitumor activity.

At the same time chemists, toxicologists, and pharmacologists have been developing other tools for using knowledge of chemical structures to characterize the toxicity of molecules. These end-points can be as diverse as carcinogenicity, mutagenicity, neurotoxicity, and reproductive and developmental toxicity.

The balance between maximizing the desired performance and function of the chemical product while ensuring that the toxicity and hazard is reduced to its lowest possible level is the goal of designing safer chemicals. Fortunately, these goals are attainable precisely because of the fact that chemists have been pursuing the methodologies for characterization and manipulation of molecular structure.

4.4.2 Why is this now possible?

This approach to the design of safer chemicals is now possible because there have been such great advances in the understanding of chemical toxicity. It has always been straightforward to measure the performance characteristics of a chemical product to see if it can perform the function that it is designed for. It has not always been easy to identify what it is about the molecule that is responsible for its toxicity. There has been a great deal of effort over the past generation to identify the mechanisms of action of substances on the body and in the environment. These mechanisms detail the exact reactions that take place in the body or environment to bring about the toxic effect. By knowing the mechanism in detail, chemists can then modify the structure so that these reactions are no longer possible, thereby reducing toxicity.

There are several basic approaches to designing safer chemicals. If a certain reaction is essential for the toxic mechanism to be carried out, a structural change could be made to ensure that the reaction could not take place. Of course, any structural modifications would ensure that the function and performance of the molecule were still preserved.

The second tier would be used in cases where the exact mechanism is not known. In these cases, a correlation may still exist between chemical structure (e.g. presence of functional groups) and the existence of a toxic effect. In this case, the functionality related to the toxic effect would be avoided, minimized, or eliminated in order to reduce or eradicate the toxic end-point.

A third tier would be effected through minimizing bioavailability. If a substance is toxic, yet cannot reach its target organ (e.g. stomach, lungs, liver), where its toxicity would be manifested, then it is rendered innocuous for all intents and purposes. Since chemists have a long tradition of knowing how to change the physical and chemical properties of a molecule, such as water solubility and polarity, they can easily manipulate molecules to make them difficult or impossible to be absorbed through biological membranes and tissues. By eliminating the absorption and bioavailability, toxicity is concurrently decreased. Therefore, as long as the change in the properties to make the molecule less bioavailable does not impair the intended function and use of the molecule, it will be both efficacious and less toxic.

4.5 The use of auxiliary substances (e.g. solvents, separation agents) should be made unnecessary wherever possible and innocuous when used

4.5.1 The general use of auxiliary substances

In the manufacture, processing, and use of chemicals, there are auxiliary substances used at every step. An auxiliary substance can be defined as one that aids in the manipulation of a chemical or chemicals, but is not an integral part of the molecule itself. The use of these substances is designed to overcome specific obstacles in the synthesis or production of a molecule or chemical product. Many auxiliary substances have come into such widespread use that there is seldom an evaluation as to whether or not they are necessary. This is true of solvents in many cases, as well as for separation agents for many operations. Often, these auxiliaries can possess properties that are of concern for human health and the environment.

4.5.2 Concerns for solvents

In the case of solvents that are commonly used, there are a number of concerns associated with them. Halogenated solvents such as methylene chloride, chloroform, perchloroethylene, and carbon tetrachloride have long been identified as suspected human carcinogens. Through a different mechanism, benzene and other aromatic hydrocarbons have also been implicated in the causation or promotion of cancer in humans and other animals. All of these substances are widely used and highly valued because of their excellent solvency properties in a wide range of applications. Those benefits, however, are coupled with the above-mentioned health hazards.

4.5.3 Environment

Concerns about solvents extend beyond the direct implications for the health of humans and wildlife into the realm of the effects on the environment and the ecosystem in which we all reside. Perhaps the most well known of the environmental implications of solvents is that of stratospheric ozone depletion. Chlorofluorocarbons (CFCs) have been widely used for a good portion of the 20th century. There is no doubting their proven effectiveness in their intended uses, from cleaning solvent, to propellant, to blowing agent for molded plastic foams to refrigeration. It is also true that CFCs have very low direct toxicity to humans and wildlife and possess low accident potential because they are both non-flammable and non-explosive. However, the effect of CFCs in depleting the ozone layer and the resulting environmental effects has been widely publicized.

Volatile organic compounds (VOCs), representing a wide range of hydrocarbons and their derivatives, have been used as solvents in a large number of applications. This class of chemicals has been implicated in atmospheric ozone generation, commonly known as smog. Through the indirect creation of this environmental effect, many individuals with respiratory problems suffer great distress.

Regulations have been constructed under the Clean Air Act in the United States, as well as a variety of other legal authorities, to control many classes of chemicals used as solvents. Because the regulatory provisions often cost significant amounts of money to comply with, many companies are looking at either alternative substances to use as solvents, or, in a more fundamental approach, alternatives to using solvents in the first place.

There also exist concerns for the auxiliary substances used in achieving separations. Most of these concerns are not centered on specific classes or types of substances, as was the case with solvents, but are more generic. An example of the characteristics of separation agents that make them of concern is the material requirements. The materials used to achieve or ease the separation of products from by-products, coproducts, impurities, or other related substances can be quite extensive as well as costly. In addition, the separation methodologies can also require a great deal of energy, through either mechanical or thermal processing mechanisms. Finally, when the separations are achieved after processing, these auxiliary materials are then part of the waste stream and in need of some sort of treatment or disposal.

One common type of separation/purification methodology is the use of recrystallization, which requires the use of both energy and/or substances that are added to change the solubility of the dissolved constituents in order to achieve distinct precipitation. The materials and energy used need to be assessed not only for their contribution to the waste stream, but also for any intrinsic hazard that they may possess.

Another extremely common type of separation used throughout chemistry is chromatography. While chromatography strictly for analysis and characterization uses only trace amounts of auxiliary substances, larger scale separations are proportionately of more significant concern. Materials associated with both the mobile and stationary phases may have some hazard concerns associated with them. Even in the cases where hazard is not a significant concern, waste and energy considerations need to be built into the evaluation of what effect the auxiliary substances are having on the overall chemical process.

4.5.4 Supercritical fluids

There are certainly alternatives to the traditional use of auxiliary substances, which are being pursued for the reasons that drive all the elements of green chemistry. Some of the alternatives to traditional organic solvents include the use of supercritical or dense phase fluids, such as supercritical carbon dioxide. This system has the advantage of not only being innocuous from a human health and environmental standpoint, but also being of use in increasing the ease of separation and selectivity.

Supercritical fluids are formed by subjecting, typically, small molecules such as carbon dioxide to the appropriate temperature and pressure to attain the critical point, which results in the molecules possessing the character of a fluid that is best described as a cross between a liquid and a gas. This fluid has the property of being a tunable solvent, meaning that the properties of the solvent can be adjusted by adjusting the parameters of temperature and pressure. This allows supercritical solvent systems to replace a variety of other solvents that may possess a hazard or be highly regulated.

4.5.5 Solventless

Solventless systems have the most obvious advantage for human health and the environment as far as a consideration of hazard is concerned. Many companies and scientists in academia are developing methods where the reagents and feedstocks serve as the solvent as well. Other systems have the reagents and feedstocks reacting in the molten state to ensure proper mixing and optimal reaction conditions. There is innovative work being carried out where reactions take place in solid surfaces such as specialized clays. All of these approaches obviate the need for the auxiliary substance (solvent) to be used in the process.

4.5.6 Aqueous

Aqueous systems have been investigated for reasons of efficacy and selectivity over the years and that work has provided an excellent basis for the use of these systems as environmentally benign solvents. It is obvious that water is arguably the most innocuous substance on Earth and is therefore the safest solvent possible. Additional considerations of methods using aqueous systems as solvents and the cost of separation of products and by-products after the manufacture are important in that there is a need to ensure that the effluent from a process does not actually contain an increased concentration of contaminants compared with what would be present with traditional solvents. The use of water as a solvent has distinct advantages that must, however, be scrutinized on a case-by-case basis for their overall environmental impact.

4.5.7 Immobilized

A major problem with many solvents in relation to human health and the environment is their ability to volatilize, and thus have a detrimental effect by exposing individuals and contaminating the air. One solution that is being investigated is the use of immobilized solvents. Immobilization has taken several forms but the goal of each is the same; to maintain the solvency of a material while making it non-volatile and unable to expose humans or the environment to the hazards of that substance. In some cases this can be done by tethering the solvent molecule to a solid support or by building the solvent molecule directly on to the backbone of a polymer. In some cases new polymer substances themselves are being developed that have solvent properties and yet do not possess the properties that would make them a hazard.

4.6 Energy requirements should be recognized for their environmental and economic impacts and should be minimized

4.6.1 Energy usage by the chemical industry

Energy generation and consumption has long been known to include a major environmental effect (Fig. 4.2). Chemistry and chemical transformations must and do play a major role in capturing and converting substances into energy as well as converting existing sources of energy into a form that is usable to society. Of course there must be an on-going commitment to making that process have a sustainable profile, unlike the current state of affairs. It should also be recognized that under present circumstances, industry uses a tremendous share of all of the energy usage in industrialized nations.

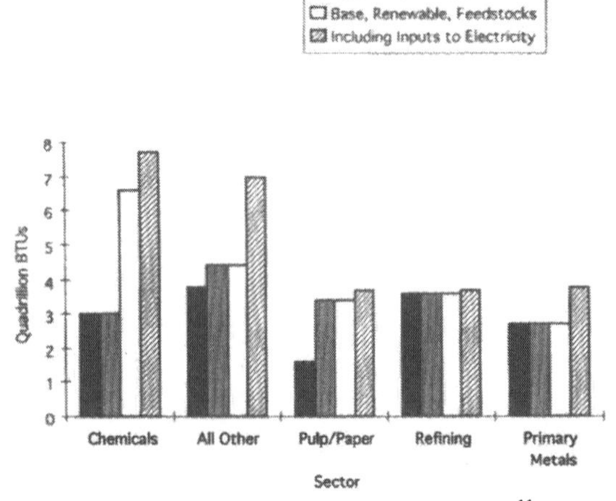

Fig. 4.2. Energy usage by the chemical industry and others.[11]

4.6.2 How energy is used

Often, there has been a prevalence for dealing with reactions that require energy input in a similar manner to one another. If the starting material and reagents dissolve well in a particular solvent, that reaction mixture has then simply been heated to reflux for the allotted time or until the reaction is complete. There is often a lack of analysis of what the heating requirements, if any, may be for a particular reaction during the design of a synthesis. It is often left to the process engineer to factor in any energy or thermal requirements in order to make the process efficient.

4.6.3 The need to accelerate reactions with heat

In cases where a reaction is being driven to its thermodynamic product, it is often the case that this will be accomplished through the use of thermal energy. This is utilized to overcome any energy of activation that needs to be traversed in order to bring the reaction to completion. One of the great advantages of catalysts is that by lowering the energy of activation needed to accomplish a particular reaction type, the amount of thermal energy that is necessary to accomplish a transformation is minimized.

4.6.4 The need to control reactivity through cooling

In some cases, reactions are so exothermic that it is necessary to control the reaction through, sometimes extensive, cooling. This thermal manner of controlling the rate of the reaction is often necessary to study reactions that are exceptionally fast and take place in microseconds. Also, in chemical manufacturing, slowing the reaction rate through cooling is sometimes necessary in order to prevent any chance of having a 'runaway' or uncontrolled reaction, which could result in a severe chemical accident. In any case, there are costs, both environmental and monetary, that are incurred with cooling, every bit as much as there are with heating.

4.6.5 Separation energy requirements

One of the most energy intensive processes in the chemical industry is the purification and separation process. Whether purification/separation be carried out through distillation, recrystallization, or ultrafiltration, energy is going to be expended in order to secure the separation of product from impurity. By designing a process that minimizes the need for separations of this kind, the chemist is also ensuring that, for the most part, there is not going to be a great deal of energy, either thermal, electric, or of other forms that will be necessary in order to obtain the product.

4.6.6 Microwaves

The use of microwave energy is a technique that is being utilized in order to effect chemical transformations rapidly, and often in the solid state, that have classically been conducted in liquid solutions.[12] In many cases the microwave techniques have shown distinct advantages in not requiring prolonged heating in order to carry out a reaction. In addition, reactions performed in the solid state also obviate the need for additional heating of all of the auxiliary solvent that is required when the chemistry is being carried out in solution.

4.6.7 Sonic

Certain transformation types (cycloadditions, pericyclic reactions) have been studied for their ability to be catalyzed by the use of ultrasonic energy in a sonicator. Through the use of this technique, the conditions of the local environment of the reacting species are changed sufficiently to promote a chemical transformation. This, like any other form of energy, would need to be evaluated for each reaction to see if it is more efficient in accomplishing the synthetic target.

4.6.8 Optimizing the reaction should mean minimizing the energy requirements

Chemists often strive to 'optimize' a reaction or entire synthetic pathway after it has been proven that it works. This use of the term optimization is usually a euphemism for striving to increase the yield or the percentage conversion of the reaction from starting materials to product. What is often not considered is the need for energy in the synthetic scheme, of the type mentioned above. It is often left to the process engineer to balance the energy requirements. However, just as in the case of the hazardous substances and hazardous waste used and generated by a reaction scheme, the chemist who designs the reaction has the greatest effect on what the energy requirements for a given synthesis or manufacturing process are. While these requirements can be adjusted and optimized, it is only through the design of the reaction system that the inherent energy requirements can be fundamentally changed. Therefore, whenever practicable, chemists should include the energy requirements of all of the various stages of the synthetic process into their evaluation process, and should strive to minimize them.

4.7 A raw material or feedstock should be renewable rather than depleting, wherever technically and economically practicable

4.7.1 What are renewable vs. depleting feedstocks?

There has been a significant focus within the scientific, industrial, and environmental communities on the use of renewable resources. The difference between a renewable resource and a depleting one can be described simply as 'time'. Depleting resources are usually thought of as the fossil fuels. Except for the fact that it would require millions of years to occur, fossil fuels could be accurately described as renewable inasmuch as vegetation could once again be converted to petroleum. Since this is not generally thought of as practical, fossil fuels are considered to be depleting. One of the true depleting resources that we have is the Sun and solar power but that again is subject to the criterion of time. Since the Sun will last millions of years, it is often referred to as an infinite supply of energy, although, unlike fossil fuels, once it is exhausted it will never be replenished.

Renewable feedstocks are most often associated with biological and plant-based starting materials. The term, however, can be equally applied to substances that are easily regenerated within time frames that are accessible to the human lifetime. Since substances such as carbon dioxide can be generated from an ubiquitous supply of sources, both natural and synthetic, CO2 can be referred to as a renewable resource. The same argument can be made with regard to methane gas, since there are a variety of natural sources, including large mammals and decomposing plant matter like marsh grass.

4.7.2 Sustainability

The most obvious concern for the extensive utilization of limited or depleting resources is the fact that, by definition, they can run out or become exhausted. This inevitability, therefore, is not regarded as sustainable either from an environmental or an economic point of view. One definition of sustainablity is the ability to maintain the development of the quality of life while not compromising the ability of our progeny to do the same. Therefore, if our generation were to consume petroleum resources to the extent that they were no longer a viable and usable option for future generations, this would violate the goals of sustainability.

4.7.3 Direct environmental effects

The environmental impact of using fossil fuels has had a pronounced effect historically on human health and the environment. There has been an environmental legacy, which includes black lung and habitat destruction related to coal mining, and oil spills and air pollution related to petroleum refining.

4.7.4 Indirect environmental effects

A less obvious effect on human health and the environment is derived from the nature of petroleum as the pre-eminent building block for the chemical industry in the last half of the 20th century. Petroleum hydrocarbons are generally in their fully reduced form and therefore require the use of oxidation chemistry to functionalize and derivatize them in order to make a variety of other useful products. Oxidation chemistry has been some of the most polluting chemistry that has contributed to the risk to human health and the environment, primarily through the use of heavy metals as the oxidizing agents, e.g. chromium.

4.7.5 Limited supply creates economic pressure

A second obvious concern when using depleting resources as opposed to renewable resources is the fact that, as supplies are depleted, the laws of supply and demand dictate that the cost of this raw material should increase. It is true that the price of a barrel of oil is roughly the same as it was 20 years ago when adjusted for inflation, but the principles of the underlying economics are indisputable. As oil supplies are consumed, economics will dictate other options for all of the petroleum products currently on the market. As petroleum prices increase, people will become increasingly driven to utilizing these feedstocks for higher value products as opposed to burning it as fuel. One of those higher value uses has been in the basic building block chemicals and in intermediates to more complex substances. Currently, the use of biological feedstocks for constructing the same higher value chemicals and intermediates is being studied and, in some cases, commercialized in an environmentally benign manner.

4.7.6 The political effects of petroleum

It has been said that if the full cost of petroleum were born by the market, the true price of petroleum products like gasoline would be many times higher than is currently the case. Some of these costs are environmental, some are military, and some are political. As an example, if the costs of protecting the petroleum shipping lanes were factored into the price of gasoline, what would the resulting price be? If the costs of environmental clean-up from oil spills or leaking underground petroleum storage tanks were directly translated into the price of a barrel of oil, what would the price be?

4.7.7 Concerns about biological feedstocks

4.7.7.1 'Seasonal supply'

Biological and renewable feedstocks are not without their concerns, both economic and environmental, as well. One economic concern is the availability of a biological feedstock when it is required. The virtue of renewable resources being generated quickly in real-time becomes a vice when a large quantity of feedstock is continuously necessary and, owing to drought or crop failure, the feedstock supply disappears. It would be unworkable for the wheels of industry to grind to a halt due to such unpredictability. (One could, however, ask the question, is this any more unpredictable than the political winds of some of the oil-producing regions of the world?)

4.7.7.2 Land/energy usage

A second concern is the amount of land usage that would be required in order to sustain the industrial products that we have become accustomed using agricultural products as the foundation. Certainly, classical crops as feedstocks require both land and extensive energy to produce, to the point where they would be impractical for use as feedstocks. It is because of this that many non-traditional biological products and processes are being developed and utilized to make renewable feedstocks feasible.

4.8 Unnecessary derivatization (blocking group, protection/deprotection, temporary modification of physical/chemical processes) should be avoided whenever possible

4.8.1 The prevalence of this practice in chemistry

As the art and science of chemical synthesis, especially organic chemical synthesis, has become more complex and the problems that it attempts to solve more challenging, methods have been developed that require significant manipulation to traverse the hurdles. Whether the hurdle is achieving stereocontrol, effecting a reaction in the presence of a labile functional group, or some other challenge, the approach that has been developed often requires the generation of a molecular modification or a derivative of the substance needed in order to carry out a particular transformation.

4.8.2 Blocking/protecting groups

One of the more commonly used techniques is the use of blocking groups. Blocking, or protecting, groups are used to protect a sensitive moiety from the conditions of the reaction, which may jeopardize the functionality if it is left unprotected. A typical example of this type of transformation would be the protection of an alcohol by making the benzyl ether in order to carry out an oxidation in another part of a molecule without affecting the alcohol. After the oxidation is completed, the alcohol can be easily regenerated through cleavage of the benzyl ether. Derivativization of this type is particularly common in the synthesis of fine chemicals, pharmaceuticals, pesticides, and certain dyes. Obviously, in the above example, benzyl chloride (a known hazard) needs to be handled and used in the generation of the material and then regenerated as waste upon deprotection.

4.8.3 Making salts, etc. for ease of processing

It is common that certain substances need to be formulated or blended with other substances in order to effect their macro or performance properties. Properties such as viscosity, dispersability, vapor pressure, polarizability, and water solubility often have to be temporarily modified in order to allow for various processing needs. These modifications can be as simple as making the salt derivative to allow for ease of processing. Again, when the functional requirement is completed, as with protecting groups, the parent compound can be regenerated easily. Obviously, this process uses material to make the substance and generates a waste in the regeneration of the original substance.

4.8.4 Adding a functional group only to replace it

When trying to design a synthetic methodology, a chemist strives for high selectivity in each reaction. When there are several reactive sites in a molecule, it is often wise to direct the reaction to the site that is desired. This can be accomplished by first derivatizing the site in a way that will make it more attractive to the reactive species and will provide a good leaving group. For example, it is common to use halogen derivatives to carry out nucleophilic substitution reactions. The halogen makes the site more attractive by making it more electropositive and the halogen itself serves as a good leaving group. Needless to say, there is then a need to form the derivative, which consumes reagents, and the reaction of the substance then generates halogen waste during the course of the desired reactions.

4.9 Catalytic reagents (as selective as possible) are superior to stoichiometric reagents

There are few reactions where reactant A and reactant B form product C in which all of the atoms contained within A and B now reside in C, and no other reagents are needed. In those rare cases, stoichiometric reactions are equally environmentally benign from the criterion of material usage as any other type of reactions. However, it is more often the case with stoichiometric reactions that:

1. one of the starting materials, A or B, is a limiting reagent and therefore there will be unreacted starting material left over even in a reaction with 100% yield;
2. one or both of the starting materials are only partially needed for the end-product and so the balance of the molecules go to the waste stream; or
3. additional reagents are needed to carry out or facilitate the reaction and those reagents will need to be discarded in the waste stream when the reaction is completed.

It is for these reasons that catalysis, where available, offers some striking advantages over typical stoichiometric reactions in many cases.

The role of catalysts is to facilitate a transformation that is desired without being consumed as part of the reaction and without being incorporated in the final product. This 'facilitation' can take several different forms, including:

Selectivity enhancement. There has been a large amount of effort and focus on the area of selectivity that has been centered on catalysis. Selective catalysis has been achieved to ensure that the degree of reaction that takes place is controlled (e.g. monoadditions vs. multiple addition), the site of reaction is controlled (e.g. C-methylations vs. O-methylations), and the stereochemistry is controlled (e.g. *R* vs. *S* enantiomer). With advances in catalytic selectivity, there are concurrent benefits for green chemistry. This is the case because both starting-material utilization is enhanced and waste production is minimized.

Energy minimization. In addition to benefits in material usage and generation, catalysis has made significant advances in the area of energy usage. By lowering the activation energy of a reaction pathway, catalytic systems not only achieve control, but also lower the temperatures that are necessary to effect a reaction. In large-scale commodity chemical processes, this energy balance issue can be the single most important factor from both an environmental and economic impact assessment point of view.

In comparing catalytic versus stoichiometric processes, the advantage of catalysis is that, while a stoichiometric reagent will generate one mole of product for every mole of reagent used, generally speaking, a catalyst will carry out thousands, if not millions, of transformations before it is exhausted.

4.10 Chemical products should be designed so that at the end of their function they do not persist in the environment and break down into innocuous degradation products

4.10.1 The current situation

A major concern with regard to chemicals in the environment is that of so-called 'persistent chemicals' or 'persistent bioaccumulators'. This means simply that after they have been discarded or otherwise released into the environment, such chemicals remain in the same form in the environment or are taken up into various plant and animal species and accumulate in their systems. Often, this accumulation is detrimental to the species concerned, either through direct effects or through indirect toxicity.

Since the use that the chemical product was designed for probably did not account for how it would effect human health and the environment after it was disposed of, there are many legacy problems to be dealt with because of the existence of chemical products that persist beyond their useful lifetime.

4.10.2 Persistence in the environment

4.10.2.1 Plastics

Plastics are one common class of compound that have historically been hailed for their durability and long life. The result, however, was that in landfills, oceans, and other aquatic media, plastics caused environmental concerns, often as a result of their physical attributes as opposed to their chemical characteristics. Sea birds ingesting indigestible plastics and the like became a problem and so there was an effort to make plastic that would break down in the environment, known as 'biodegradable' plastic.

4.10.2.2 Pesticides

Many pesticides are organohalogen-based structures. These chemicals, whilst efficacious, also tend to bioaccumulate in many plant and animal species, often in the adipose tissue, or fat cells. This can cause damage both to the species itself, or, if that species is consumed, to humans. The pesticide DDT was one of the first pesticides of this type to be shown to exhibit a wide range of behaviors that were detrimental in this way.

4.10.2.3 Just as you design for function, consider degradation as a function

Whenever a chemical product is being designed, the first thing that is done is an assessment of the properties that the substance needs to possess. In approaching the problem of persistent bioaccumulators, the green chemistry considerations would be to address the disposition of the substance after its function is completed as well as designing the substance to achieve its primary function. If a plastic needs to serve as a garbage bag for example, it needs to possess certain properties; and, in addition to those properties, it should be designed so that it does not remain in its intact physical state in the environment after its useful life is over.

4.10.2.4 Designing for biodegradability

In designing a chemical for biodegradability, one must assess what substances the parent product will break down into. It is possible to place features and functional groups in the molecular structure of a chemical product that will facilitate its disassembly. Functionality, which is susceptible to hydrolysis, photolysis, or other cleavage, has been used successfully to ensure that products will biodegrade.

It is equally important, however, to recognize that the degradation products themselves may possess toxicity or other hazards that must be assessed. It is obvious that if a products is designed to break down into substances that pose an increased risk to human health and the environment, then the green chemistry goal has not been accomplished. Just as with any other type of green chemistry product or process, biodegradation processes should include the effects on human health, ecosystems, wildlife, and the overall pollution load.

4.11 Analytical methodologies need to be further developed to allow for real-time, in-process monitoring, and control prior to the formation of hazardous substances

Analytical chemists have characterized and detected environmental problems since the beginning of the environmental movement. An area of focus within the analytical community now is to develop methods and technology that allow the prevention and minimization of the generation of hazardous substances in chemical processes.

The development of process analytical chemistry for green chemistry is based on the premise that 'You cannot control what you cannot measure'. In order to effect changes on processes during their operation, you need to have accurate and reliable sensors, monitors, and analytical techniques to assess the hazards that are present in the process stream.

In order to accomplish the goals of green chemistry, the analytical techniques that are being developed can be used both in-process and in real-time. Using these features a chemical process can be monitored for the generation of hazardous by-products and side reactions. When these toxic substances are detected at even the smallest trace levels it may be possible to adjust the parameters of the process to reduce or eliminate the formation of these substance. If the sensors are interfaced directly with process controls, this hazard minimization may very well be automated.

Another example of the use of process analytical chemistry is in the monitoring of the progress of reactions to determine their completion. In many cases, chemical processes require the continuous addition of reagent until the reaction is complete. If there is a real-time, in-process monitor to allow determination of completion, then the need for additional excess reagent can be obviated and potentially hazardous substances can be eliminated from use and will not find their way to the waste stream.

4.12 Substances and the form of a substance used in a chemical process should be chosen so as to minimize the potential for chemical accidents, including releases, explosions, and fires

The importance of accident prevention in chemistry and the chemical industry cannot be overstated. There have been a number of notable chemical accidents that have resulted in the mobilization of public opinion to control the use of chemicals. The accidents in Bhopal, India, and Seveso, Italy, and others, have resulted in the loss of hundreds of human lives. The hazards posed by toxicity, explosivity, and flammability all need to be addressed in the design of chemical products and processes. The goals of green chemistry must involve the full range of hazards and not be focused simply on pollution or ecotoxicity.

It is possible to increase accident potential inadvertently while minimizing waste generation in preventing pollution. In some cases where the recycling of a solvent from a process may have advantages from the perspective of pollution prevention and release to the environment, it may also increase the potential for a chemical accident or fire. A process must balance the desire for pollution prevention with that for accident prevention.

Approaches to the design of inherently safer chemistry can include the use of solids or low vapor pressure substances in place of the volatile liquids or gases that are associated with the majority of chemical accidents. Other approaches include avoiding the use of molecular halogens in large quantity by substituting reagents that carry the halogens to be transferred in a more innocuous manner.

The utilization of 'just-in-time' techniques involves the generation and rapid consumption of hazardous substances within a contained process. By employing this technology, a chemical company can eliminate the need for large inventory stockpiles of hazardous substances which can pose a significant accident risk.

5 *Evaluating the effects of chemistry*

5.1 How does a chemist evaluate a chemical product or process for its effect on human health and the environment?

It is easiest to assess a chemical product or process in stages rather than trying to evaluate the whole system in its entirety. Certainly, with the number of parameters that it would be necessary to evaluate at each step of the process, it would be virtually impossible to conduct a systematic assessment on a product or process as a whole. Therefore, a simple approach is to identify the various parts of the process and to ask the necessary evaluative questions at each of those stages. As was mentioned earlier, the five aspects can be thought of as follows.

1. Starting materials/feedstocks
2. Reaction types
3. Reagents
4. Solvents and reaction conditions
5. Chemical products/target molecules

For each of these aspects, the same questions need to be addressed. Whether dealing with a solvent, reagent, product, etc., there are essential characteristics that need to be known in order to conduct a green chemistry evaluation.

5.1.1 Toxicity to humans

Once information on the origin of the substance has been evaluated, we must then address the question of what is the hazard of the substance itself. As mentioned earlier, there can be no risk without hazard and therefore the question of hazard is central to conducting a green chemistry evaluation. It is commonly the case that the toxicity of a substance to humans is often addressed separately from that of the toxicity to other species in the environment when conducting a risk assessment. The reasons for this are likely to be both anthropocentric as well as scientific. Scientifically, there are cases in which the toxicity of a substance will differ widely between species and therefore extrapolations are unreliable. In these cases it is wise to deal with the toxicity to humans separately within the evaluation.

The toxicity of substances to humans has been the focus of extensive testing and has resulted in a number of large databases. The databases contain voluminous information on the levels at which substances will produce a toxic effect in humans. This factor, or potency, is a marker of preferability of one substance versus another (Fig. 5.1). The nature of the effect can be wide ranging, from irritation (e.g. of the skin) to cancer. It is for this reason that one needs to consider not only the potency of the substance in creating a toxic effect but also the severity of the end-point (Fig. 5.2).

In addition to the severity and the potency of an effect, the reversibility or irreversibility of a given effect must be considered. In many cases a toxic effect can be severe as well as potent and yet be totally reversible. In contrast, other substances are known to possess end-points that are not considered as severe and which are induced only at very high levels of exposure, yet are totally irreversible. Therefore, the evaluation of human toxicity has yet another parameter, reversibility, which must enter into the evaluation process (Fig. 5.3).

Therefore, the three factors that must be considered in the evaluation of toxicity to humans are severity, potency, and reversibility. Each one of these factors can greatly enhance or mitigate concern for the hazard of a particular chemical substance.

Chemical Substance X:

The substance possesses no other hazards except for blindness in humans. When exposed to levels exceeding 10 parts per billion, humans will lose all eyesight.

Chemical Substance Y:

The substance possesses no other hazards except for blindness in humans. When exposed to levels exceeding one million parts per billion, humans will lose all eyesight.

Green Chemistry Evaluation:

Chemical X and Chemical Y cause the same toxic effect; blindness in humans. Chemical Y is preferable because it is one hundred thousand times less potent than Chemical X in causing blindness. All other factors being equal, Chemical Y is evaluated as being preferable to Chemical X in the green chemistry evaluation.

Fig. 5.1. An example of a comparison of potency.

Chemical Substance X:

This substance possesses toxicity only to humans. The only toxic end-point that it possesses is at 100 parts per million, at which it acts as a lachrymator or tearing agent. Tearing continues as long as the person is exposed to chemical substance X at 100 parts per million or above.

Chemical Substance Y:

This substance possesses toxicity only to humans. The only toxic end-point that it possesses is at 100 parts per million, at which it causes neurological and central nervous system damage. The neurological damage continues as long as the person is exposed to chemical substance Y at 100 parts per million or above.

Green Chemistry Evaluation:

The potency of Chemical X and Chemical Y are identical. The difference is the end-point. Because neurological/central nervous system damage are more severe end-points than tearing/lachrymation, Chemical Substance X would be preferable to Chemical Substance Y in a green chemical evaluation where all other factors are equal.

Fig. 5.2. An example of a comparison of severity.

Chemical Substance X:

This substance possesses toxicity only to humans. It causes severe respiratory distress at levels of 10 parts per million or above. All respiratory distress symptoms disappear when the substance is removed or is present at less than 10 parts per million.

Chemical Substance Y:

This substance possesses toxicity only to humans. It causes severe respiratory distress at levels of 10 parts per million or above. All respiratory distress symptoms remain even after the substance is removed or is present at less than 10 parts per million. Damage to the respiratory tract is permanent.

Green Chemistry Evaluation:

Both Chemical X and Chemical Y are equally potent and share an identical end-point of identical severity. The only distinction is the reversibility. Chemical Substance X is certainly preferable than Chemical Substance Y (all other factors being equal) from a green chemistry evaluation standpoint because of the reversibility of the toxic effect.

Fig. 5.3. An example of a comparison of reversibility of effect.

5.1.2 Toxicity to wildlife

In conducting a green chemistry evaluation, one must consider the effects of a substance on wildlife as well as on humans. The obvious difficulty with this is that while humans are one species, 'wildlife' covers an enormous number of species and any evaluation could not hope to include the full range. None the less, there are sources of data that can be used to incorporate the effect of a chemical substance on wildlife in general, through extrapolations.

Because of obvious logistical considerations, the vast majority of toxicological data for various substances is based on animal testing. This provides a very large database on how thousands of chemicals will effect various species for all types of toxic end-points. In addition to these empirical databases, there has been the development and continuing enhancement of techniques known as structure–activity models. These models allow for the extrapolation of data from chemicals that have been tested to chemicals that have not yet been tested on living systems. This is done through a series of analogies where a compound that has no data is linked to a substance for which there is toxicological data and to which it is very closely related. This relationship is usually structural although other factors such as electronic configurations, molecular volume, and other parameters may also play an important role. By knowing the toxicity of a substance of a very close analogue one can often predict the approximate toxicity of the related substance, at least to identify the end-points of concern and their potency.

By using these data sources, one can assess the effect of a chemical on wildlife, or of a particular substance on a particular species. The choices resulting from this information can sometimes be problematic. What does one do in the case where substance A is hazardous to one particular species and substance B is hazardous to a different species? This would require a more comprehensive risk assessment to identify if it was likely that the species in question would have a reasonable likelihood of being exposed to the substance in question.

Another problematical situation is where a substance may actually be beneficial to one species and detrimental to another. Such is the case with phosphates in certain aquatic environments. In the right circumstances, phosphates can be a nutrient to various types of algae resulting in 'algal blooms' or an over-abundance of algae. This process can result in the deoxygenation of the water and has been known to cause large kills across many species of fish. This type of indirect toxicity is another factor to consider when evaluating the effect on wildlife. There will be further discussion of indirect toxicity in the section on environmental effects.

5.1.3 Effects on the local environment

The use of certain substances is known to cause changes in the environment that can be detrimental. While the substances themselves may show no toxicity to human health or the health of wildlife, the overall effect from the changes in the environment can be quite significant. The changes known to be caused by chemicals can range from the effects on a particular pond or stream to effects on the planet as a whole.

An example of an effect to the local environment is that of acid rain. As is now widely known, many of the by-products of combustion processes, such as the oxides of nitrogen and sulfur, are washed out of the atmosphere and form acid precipitation. This acidic rain or snowfall has been known to cause widespread deaths of aquatic species and plants, especially in smaller ponds and lakes. The oxides themselves, if tested for toxicity to the species that are ultimately effected, would, in the same instances and concentrations, be regarded as safe for these end-points. It is the indirect toxicity caused by changes in the environment that are responsible for the flora and fauna deaths. This example illustrates why one must consider any changes in the local environment where a substance is being used in order to characterize fully the hazard that it may pose.

A second example of changes to the local environment, is that of atmospheric ozone build-up. Certain substances, such as hydrocarbons, can trigger the formation of ozone in the atmosphere, commonly known as smog. This ozone can result in respiratory distress for sensitive populations as well as cause deterioration in visibility and the general quality of life. Again, if the hydrocarbons were tested for these end-points, the results would be negative because it is the indirect consequences of the hydrocarbons in the atmosphere and the resultant change in the local environment that causes the adverse effects.

5.1.4 Global environmental effects

The environmental effects of certain substances can affect the whole planet and all of the living things on it. Any green chemistry evaluation must incorporate the global environmental effects in order to have a characterization that is as complete as possible. Two of the most widely known global effects, which result from the use and generation of certain substances, are global climate change and stratospheric ozone depletion.

Global climate change, otherwise known as the greenhouse effect or global warming, results from the generation of carbon dioxide and other gases in the atmosphere. As the concentration of these so-called greenhouse gases increases in the atmosphere, the global temperature increases, and at low concentrations of these gases, the global temperature decreases. This is a natural phenomenon that has taken place over geological time. With the increased generation of carbon dioxide from anthropogenic sources, such as the combustion of fossil fuels, there is concern that there may be an increase in carbon dioxide in the atmosphere and that this could result in an increase in the average temperature of the planet. This increase, it is proposed, could be catastrophic if left unchecked. Again, carbon dioxide itself is considered innocuous to all species and is obviously necessary for plant life, yet the indirect effects that could result from a build-up of CO2 in the atmosphere could be very significant.

Stratospheric ozone depletion has been the cause of a great deal of concern and the 1995 Nobel Prize in Chemistry was awarded for the elucidation of this problem. At the center of this issue is a class of substances, chlorofluorocarbons or CFCs, that have been implicated in causing the depletion of ozone levels in the stratosphere. Since the ozone layer protects the Earth from many of the Sun's harmful rays, this ozone depletion could result in toxic effects such as skin cancer and other maladies. CFCs, when introduced in the first half of this century, were extolled for their superior refrigeration properties coupled with extremely low toxicity to humans. It was only decades later that the ozone depleting properties of CFCs were characterized.

This example once again illustrates that although a substance, in this case CFCs, may not be directly toxic to humans or wildlife, its effect on the global environment could result in indirect hazards to all life on the planet. An additional point, which the case of stratospheric ozone depletion illustrates, is that knowledge of the effects of a chemical substance can change greatly over time.

Conducting a green chemistry evaluation is the same as any other evaluative exercise and can only be expected to use the best available data. While a scientist can not predict how increased future knowledge may change the current understanding of the effects of certain substances, they can incorporate the best available current knowledge in the decision-making process. Historically, the decision to use a particular chemical versus an alternative never considered the effects on the planet, the local environment, human health, or wildlife, so this is definitely a step in the right direction.

6 *Evaluating feedstocks and starting materials*

One of the greatest effects that the manufacture, processing, and use of a chemical substance has can be determined from its starting material. If the starting material, or feedstock from which the chemical product is made, has a negative environmental effect, it is very likely that the overall effect of the chemical substance itself will have a net negative effect. It is for this reason that evaluation of the feedstock or starting material is fundamental when conducting a green chemistry assessment of a chemical product or process.

The magnitude of the effect of the feedstock on the overall profile of the chemical product is dependent on a number of parameters, including the complexity and the length of the process that makes the chemical. If the preparation of the chemical is a one-step catalytic conversion to a petrochemical building block, then the environmental profile of the starting material is of paramount importance. If the end-product is a pharmaceutical that is manufactured via a 12-step synthetic pathway with complex processing and purification, the importance of the starting material may be somewhat diminished. In either of these cases, the profile of the starting material is the first step in evaluating the green chemistry of the chemical product or process.

6.1 Origins of the feedstock/starting materials

The first question is 'What is the origin of the feedstock/starting material in question?'. Is it mined, refined, synthesized, distilled, etc.? What are the consequences of the origins of the substance? A chemical that originates from the use of an otherwise useless waste product that would need to be disposed of may very well have distinct environmental advantages to its use. A chemical that originates from a process that depletes a limited natural resource or results in irreversible environmental damage may have very negative effects on the environment. Both of these situations could be true regardless of whether the actual substance in question is harmful or innocuous. It is for this reason that one must first ask the question, how did this substance originate?

6.2 A renewable or a depleting resource

The information that derives from the question of what is the origin of the feedstock or starting material can be used to answer a second question in the green chemistry evaluation; 'Is the feedstock renewable or is it depleting a limited resource?'. One could argue that all substances are renewable given long enough time. For the purposes of a green chemistry evaluation, the distinction between depleting versus renewable should be placed in the context of a time frame of human experience, such as an average human lifespan rather than of geological time. As such, it is reasonable to consider that petroleum and other fossil fuel-based feedstocks are depleting and those feedstocks based on biomass and agricultural wastes are renewable.

It is possible that the same feedstock can come from a renewable or a depleting source, depending on its origin. In the case where carbon dioxide is used as a carbon-source building block, one could make the argument that if the CO2 were made from burning fossil fuels then it would be considered depleting whereas if it were generated from the combustion of biomass it might be considered renewable. In many cases, this argument is moot since the carbon dioxide is often generated from carbonate mineral deposits.

In the analysis of feedstocks, it would be preferable to have a sustainable supply, not only for current generations, but also for posterity. This factor of depleting resources is of concern not only for environmental reasons but for economic reasons as well, since a depleting resource will certainly increase the costs of manufacture and the price of purchase for the products being produced. Therefore, a renewable starting material would be preferable to one that depletes the natural resources of the Earth, if all other factors are equal.

6.3 Hazardous or innocuous feedstock

An important question that will be at the heart of each step of a green chemistry evaluation is the consideration of intrinsic hazard to human health and the environment. A feedstock must be evaluated to determine whether it possesses chronic toxicity, carcinogenicity, ecotoxicity, etc. It is the feedstock that is going to have to be produced, often in large quantities, to manufacture a chemical product. It is the feedstock that is going to need to be handled by workers, again in large quantities, throughout the process. If this starting material does possess a significant hazard to human health and the environment, its effect will be felt throughout the life cycle of the chemical product.

6.4 Downstream implications of the choice of feedstock

The decision of what feedstock to use in the manufacture of a chemical product can and does have implications well beyond those directly attributable to the starting material itself. If the choice of feedstock will require that an extremely toxic substance is needed as a reagent to complete the transformation in the next step in the synthetic pathway, then it was the choice of feedstock that indirectly caused an even greater negative environmental impact than might be assumed on a first analysis. It could be true that an innocuous, renewable feedstock whose origins did not cause environmental damage, might still cause damage to human health and the environment because of the downstream substances that are requisite upon its utilization. It is therefore essential, as always when conducting a green chemistry analysis, not only to focus on the substances being analyzed at any particular part of the evaluation, but also to consider the implications and indirect consequences of using those materials.

7 *Evaluating reaction types*

When considering a synthetic procedure, it is useful to identify the general type of transformation being employed. Several constructive strategies based on well-known and well-documented routes are often available. The synthetic chemist is trained to identify these pathways. A thorough literature search of the massive collection of chemical books, journals, and other publications will uncover any prior reports of a given synthetic process, or at least give information concerning similar systems that have been investigated in the past. If the particular transformation is unprecedented, the synthetic chemist then draws on their own ingenuity, expertise, and knowledge of basic principles of what transformations are theoretically possible to design or create a new pathway that leads to the desired target material. The information available to the researcher, in combination with his or her own chemical skills, will lend insight into a synthetic sequence's likely efficiency. A fundamental understanding of mechanistic organic chemistry allows the practising chemist to predict the specificity of a reaction process and thus speculate and in some cases predict the relative success of a given synthetic route.

7.1 What are the different general types of chemical transformation?

Because the success of a given transformation has been historically characterized by the quantitative amount of material transformed from reactant or starting material into desired product, most, if not all, of the chemical literature and educational training of a synthetic chemist is focused on recognizing and critically assessing a synthetic sequence's yield and specificity. This section provides some illustrative examples of how one may enhance that evaluation process of a given synthetic scheme beyond these traditional measures to predict some environmental consequences and use this information to design green chemistry alternatives.

The various reactive processes available to the synthetic chemist can be categorized in a variety of different ways. Some categorizing schemes classify the synthetic procedures according to which functional groups (specific atomic sequences within a molecule) are being constructed or manipulated. Other schemes focus on the number of atoms being transformed during the specific synthetic transformation. For the purpose of evaluating which strategies may be considered more environmentally benign it is useful to identify the general type of reaction that is occurring in the particular sequence. In this way, transformations may be viewed as addition reactions, substitution reactions, elimination reactions, pericyclic reactions, and oxidation/reduction reactions. It is important to point out that this classification scheme is somewhat arbitrary. The intention here is to present an illustrative approach as to how one evaluates a synthesis with regard to its environmental consequences. It is not desirable to generate an all-inclusive listing of every reaction type. As will become obvious during the following discussion, generalizations can be limited in their usefulness. While each reaction type is representatively similar, the precise nature of the specific starting material and by-products will be of overriding importance when evaluating the environmental aspects of a given synthetic sequence.

7.1.1 Rearrangements

Rearrangement reactions, as the name indicates, are classes of reactions where the atoms that comprise a molecule change their orientation relative to one another, their connectivity, their bonding pattern, etc. to yield a new molecule. These rearrangement reactions can be effected through a variety of methodologies including thermal, photo, and chemical induction. A primary characteristic of this reaction class from a green chemistry analysis perspective is that the feedstock, or starting material, and the end-product, or target molecule, both contain the same atoms. Therefore, there is no intrinsic waste generated necessarily from a rearrangement reaction. From an inherent analysis, a rearrangement reaction is fully atom economical and fully efficient.

As will be true of all the reaction types, the specific efficiency of actually conducting a rearrangement, e.g. What is the conversion rate to product? Are there side reactions? What energy input is required?, has to be evealuated on an empirical basis.

7.1.2 Addition reactions

Addition reactions, as illustrated in Fig. 7.1, are of widespread utility. In this general synthetic scheme, a multiply bonded, or unsaturated, substrate is treated with a reactive reagent. Heterolytic or homolytic addition to the termini of the unsaturated moiety gives rise to a product that has incorporated new functionalities. The addition of bromine to an olefin, ethylmagnesium bromide to a carbonyl, or hydrogen cyanide to an α,b-unsaturated carbonyl compound serve as examples of addition reactions (Fig. 7.2).

Fig. 7.1. Schematic representation of an addition reaction.

These schemes represent chemical reactions that are self-contained. Notice that all reagents are consumed during the reaction sequences. There is an equimolar quantity of reacting species which gives rise to a single molar quantity of product. No additional by-products are generated. The efficiency of this reaction type can be quite high, provided both added components are desired in the final product. As is often the case with any synthetic strategy, further manipulation of the added functionalities may be necessary. When regiospecificity and chemospecificity allow for a reasonably high yield, this can be an excellent synthetic strategy. The removal, containment, recycling, or disposal of any planned by-products is not necessary, because no such by-products are produced.

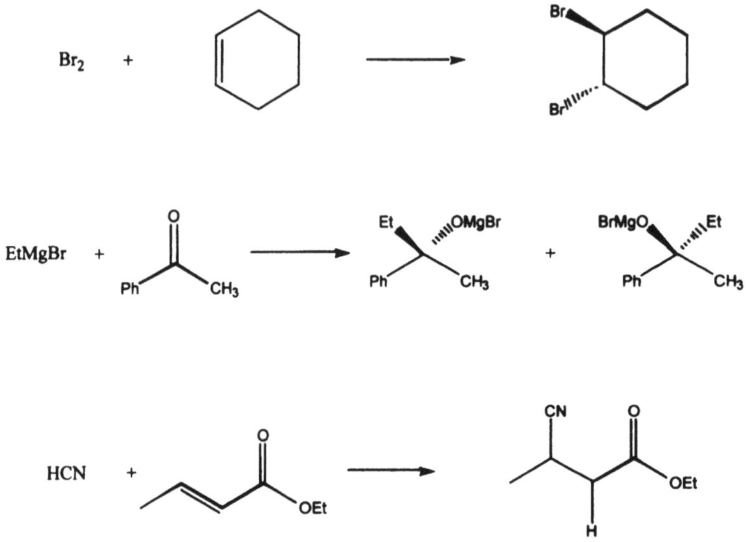

Fig. 7.2. Examples of addition reactions.

Of course, unplanned by-products that are formed during the process through uncontrollable side reactions may have to be dealt with. These side reaction by-products lead to lack of specificity and yield. Costly reactor design and engineering would therefore be required to minimize this inefficiency and, as described earlier, is the basis for much of the chemical literature on addition reactions.

Many chemical reactions, to a certain degree, will generate unplanned by-products. Purification schemes, such as recrystallization, distillation, chromatography, or other processes, will often have to be employed. The key to green chemistry design is to minimize the need for these techniques. The benefit of an addition reaction is that no planned by-product is generated.

7.1.3 Substitution reactions

Substitution reactions involve the modification of a substrate by the replacement of one functional group with another (Fig. 7.3). Classically familiar SN1 and SN2 reactions provide simple examples of this strategy (Fig. 7.4). In these cases, nucleophilic reagents displace a leaving group on an aliphatic carbon atom. The new product incorporates the nulceophile, and the leaving group is removed.

A—B + C—D ⟶ A—C + B—D

Fig. 7.3. Schematic representation of a substitution reaction.

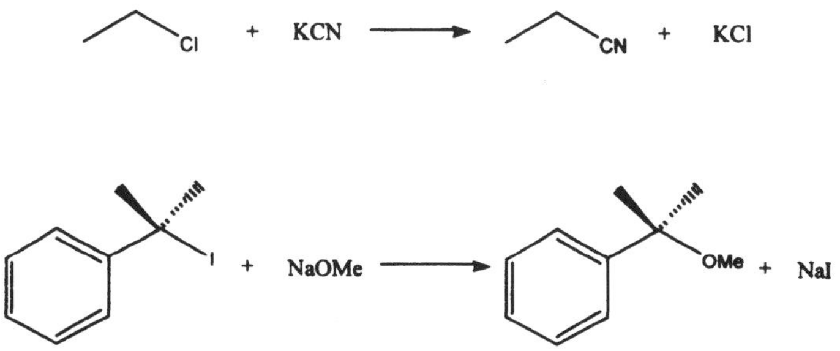

Fig. 7.4. Examples of SN1 and SN2 substitution reactions.

In most cases these syntheses focus on the reagent that is undergoing nucleophilic attack as the desired product. It must be pointed out that there exist examples where, in fact, the leaving group is the desired product. Various synthetic strategies incorporating protecting groups can be used that isolate the leaving group itself as the compound of interest. Potassium iodide demethylation of a carboxylic acid methyl ester to give the free carboxylate salt (Fig. 7.5) is illustrative of this concept.

More complicated reactions, often involving unsaturated substrates, combine addition and elimination reactions (of both possible linear sequential orders) to effect net substitution reactions.

Electrophilic aromatic substitution and elimination/addition reactions are representative of these types of transformations (Fig. 7.6).

Fig. 7.5. Use of a protecting group to isolate the leaving group.

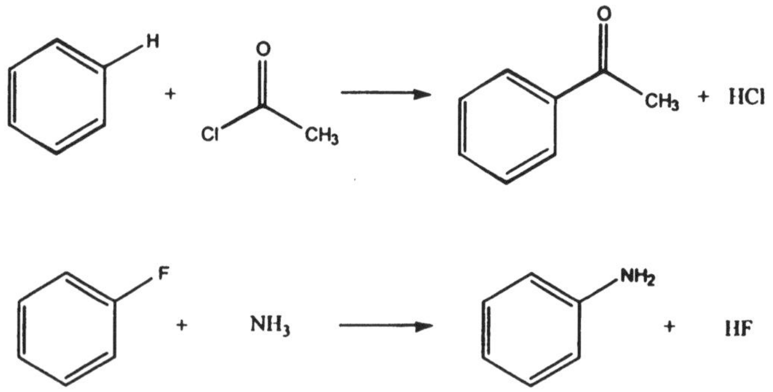

Fig. 7.6. Electrophilic aromatic substitution and elimination/addition reactions.

Obviously the environmental properties and nature of the leaving group generated will determine the appropriateness of this methodology. Halides, esters, alcohols, and inorganic derivatives that serve to activate the leaving group can be designed with hazard-related consequences in mind. If a substitution reaction sequence can be designed where the leaving group has been carefully selected, this pathway can be both convenient and efficient. The unfortunate distinction of this reaction type is that the generation of the synthetic by-product is unavoidable and a direct consequence of the synthetic design. From a thermodynamic perspective there is usually no unfavorable entropic relationship for these reactions. The number of reacting species is equal to the number of species produced. When considering the engineering consequences of large-size/scale conditions, this entropic condition can have some importance.

Because there exists a planned by-product from this reaction sequence, accommodations for the extra compound produced must be made. Efforts involving reagent containment, recycling, and redistribution, in addition to general research in the area of industrial ecology have attempted to offset much of the negative appeal of by-product-producing reactions. Where avoidance of the inherent waste generation is not possible, these avenues of by-product compensation are available to the synthetic chemist to minimize the losses inherent in these pathways.

An additional consideration for this reaction strategy is the requirement for a catalyst. While not always necessary, Lewis acids or bases are incorporated into the synthetic procedure to accelerate reaction rates or direct chemospecificity, regiospecificity, or, in some cases, stereospecificity. The effect of these reagents, when used, must not be overlooked. If the auxiliary reagent is used in substoichiometric quantities, the effect can be somewhat mitigated, but must still be accounted for in the final analysis. The use of catalytic reagents is dealt with in more detail later.

7.1.4 Elimination reactions

Elimination reactions are important avenues to generate unsaturation within molecules (Fig. 7.7). Schematically, an elimination reaction is the reverse of an addition reaction. The elimination scheme generally follows the heterolytic or homolytic dissociation of molecular components on adjacent atoms, increasing the bond order between these atoms. Dehydration of an alcohol to generate an olefin and loss of an alcohol from a hemiacetal to give an aldehyde are examples of elimination reactions (Fig. 7.8).

Fig. 7.7. Schematic representation of an elimination reaction.

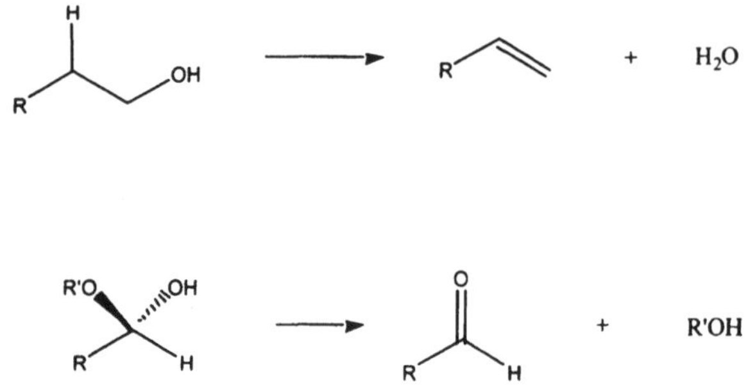

Fig. 7.8. Examples of elimination reactions.

As with substitution reactions, eliminations necessarily involve the generation of a leaving group, except in some cases where the leaving group is intramolecularly attached (Fig. 7.9). The general applicability of this type of molecular rearrangement is necessarily limited because the strategic placement of two functional groups is required for successful implementation. Nevertheless, if such a strategy can be incorporated into a synthetic design, the benefits are obvious.

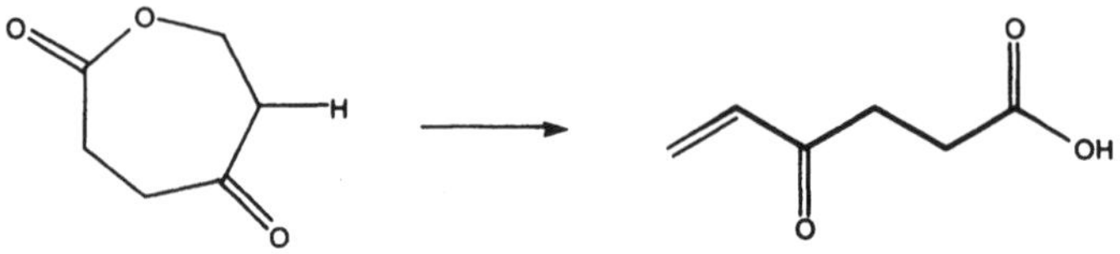

Fig. 7.9. An example of an elimination reaction where the leaving group is intramolecularly attached. This results in a molecular rearrangement.

As in the case of substitution reactions, when a leaving group is produced as a consequence of the reaction design, the environmental implications of the leaving species must be evaluated and controlled. The desired target compound, while usually the species undergoing the loss of the molecular by-product, can sometimes be the leaving group itself.

7.1.5 Pericyclic reactions

Pericyclic reactions are synthetic strategies governed by frontier molecular orbitals. As originally described by Woodward and

Hoffman, these procedures provide the synthetic chemist with a diverse array of reaction pathways. Diels–Alder reactions, 1,3-dipolarcycloadditions and [3,3']-sigmatropic rearrangements are representative bond-forming pericyclic reactions (Fig. 7.10).

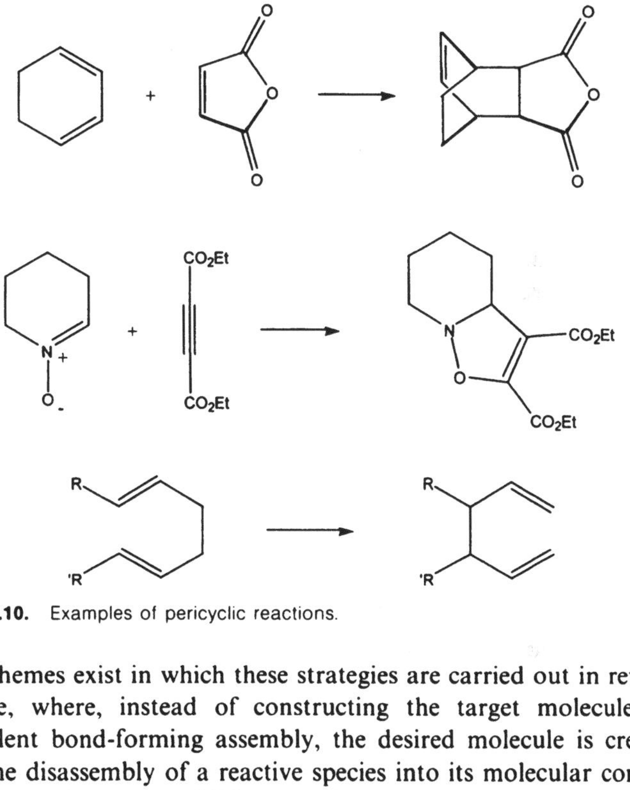

Fig. 7.10. Examples of pericyclic reactions.

Schemes exist in which these strategies are carried out in reverse mode, where, instead of constructing the target molecule via covalent bond-forming assembly, the desired molecule is created by the disassembly of a reactive species into its molecular components. Analogous to addition vs. elimination reactions, bond-forming pericyclic reactions are often more easily implemented in the environmental framework than bond-breaking pericyclic reactions. Controlling the produced by-product, however, allows for the use of these strategies.

In the same way that some substitution reactions involve a combination of addition and elimination processes within the synthetic design, combinations of pericyclic bond-forming and bond-breaking sequences are available. The Diels–Alder reaction of an acetylene with a 1,2,4-triazine illustrates this concept (Fig. 7.11). In this reaction, the acetylene first forms the Diels–Alder adduct with the triazine. The intermediate then collapses with loss of elemental nitrogen to generate a new pyridine molecule.

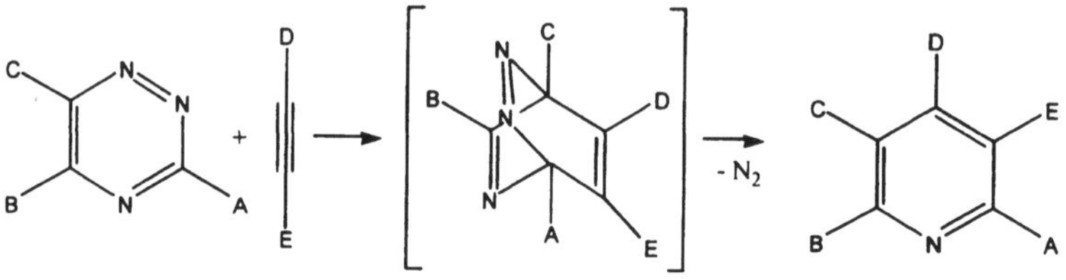

Fig. 7.11. A Diels–Alder reaction illustrating pericyclic bond forming and bond breaking.

7.1.6 Oxidation/reduction reactions

The control and manipulation of the oxidation state of a molecular species can be achieved both chemically and electrochemically. The oxidation of methane to methanol, methanol to formaldehyde, formaldehyde to formic acid, and formic acid to carbon dioxide is a representative oxidation sequence (Fig. 7.12). The reverse sequence; carbon dioxide to formic acid, formic acid to formaldehyde, formaldehyde to methanol, and methanol to methane is the corresponding reductive series. The oxidation of ethane to ethylene, ethylene to acetylene, and the corresponding reverse sequence is another representative oxidation/reduction sequence (Fig. 7.13).

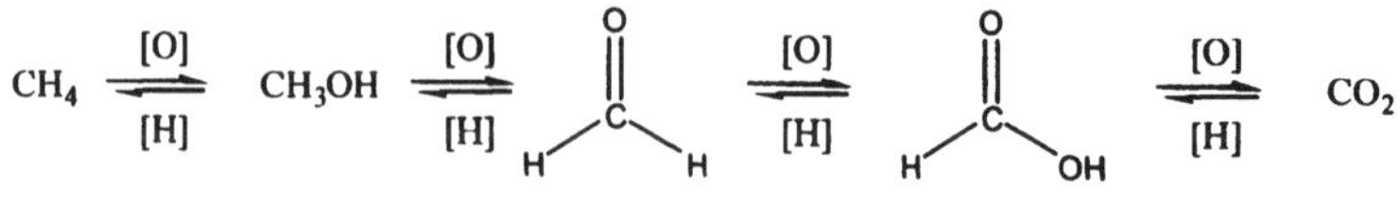

$$CH_4 \underset{[H]}{\overset{[O]}{\rightleftharpoons}} CH_3OH \underset{[H]}{\overset{[O]}{\rightleftharpoons}} HC(=O)H \underset{[H]}{\overset{[O]}{\rightleftharpoons}} HC(=O)OH \underset{[H]}{\overset{[O]}{\rightleftharpoons}} CO_2$$

Fig. 7.12. Example of an oxidation/reduction reaction.

$$H_3C{-}CH_3 \underset{[H]}{\overset{[O]}{\rightleftharpoons}} H_2C{=}CH_2 \underset{[H]}{\overset{[O]}{\rightleftharpoons}} HC{\equiv}CH$$

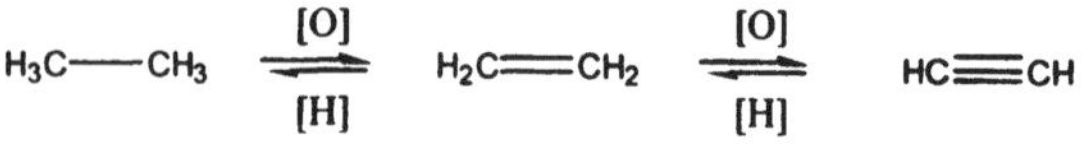

Fig. 7.13. Example of an oxidation/reduction reaction.

Chemical oxidation/reduction is when a 'courier' molecule functions to shuttle the accepted or extracted electron from the molecule being reduced or oxidized. Electrochemical oxidation/reduction involves the direct administration of electrical current to effect the removal (oxidation) or addition (reduction) of electrons to or from the target molecule.

While trade-offs certainly exist, a comparison between these two methods leads one to consider electrochemical reactions as being more environmentally benign. Many stoichiometric redox reagents are inherently toxic and their use must be carefully monitored and controlled. Unfortunately, alternative, benign reagents are not extensively available to accomplish chemical oxidation/reduction.

7.2 What is the intrinsic nature of the various reaction types?

Having characterized the general reaction type of a particular synthetic sequence as an addition, substitution, elimination, pericyclic, or oxidation/reduction reaction, a deeper evaluation of the specific chemical transformations needs to be performed. Questions of auxiliary reagents, waste generation, mass consumption, and atom economy need to be addressed.

7.2.1 Do they require additional chemicals?

Obviously a substitution reaction will require both the functional molecule being added as well as the molecule undergoing the substitution itself. In addition to the fundamental reaction profile, however, the question of auxiliary reagents must be addressed.

Chemical catalysis is often employed as a method of accelerating the rate of a reaction. By combining with one or more reactants or intermediates, a catalyst lowers the activation energy of a particular transformation by making the transition state more energetically accessible. These transition state catalysts can be applied in a stoichiometric amount (in molar quantities equal to the reacting species) or in substoichiometric amounts. When less than molar equivalents of catalyst are being used, quantities slightly lower than the reacting species or several orders of magnitude less than the reacting species can be applied. The specific amount of catalysts used is often established empirically by process engineering and reaction design.

After the stoichiometry of an auxiliary reagent is determined, an evaluation of its environmental effects can be better assessed. If the auxiliary reagent is being used in molar equivalent quantities then it should be treated differently than if it is being used in much smaller quantities. Further generalizations are not typically useful. The nature and toxicology of the specific auxiliary reagents being used will dictate the environmental implications of these synthetic strategies. In some cases a ranking will be obvious when comparing two alternative synthetic sequences. The stoichiometric use of a particularly toxic reagent will certainly be less desirable than the substoichiometric use of an environmentally benign catalyst. Often, however, the choice is not as straight forward. The substoichiometric alternative may be more toxic than the stoichiometric pathway. In this case the trade-off of toxicity and quantity must be considered.

7.2.2 Do they necessarily generate waste?

While developing a synthetic sequence it is useful to assess the material that is being generated. It is likely that the majority of the reaction product is the compound being synthesized. There are, however, other by-products that will inevitably be produced. Our previous discussions have referred to these by-products as planned or unplanned. This wording is perhaps a bit obscure. There should be no unexpected by-products! Within the stoichiometric description of a chemical reaction some materials are produced as a fundamental consequence of a reaction. Substitution reactions and elimination reactions are examples of this. Bond-breaking processes in the sequence give rise to more than one product. In a typical metathesis reaction (Fig. 7.14) one of the product materials will be the desired compound, the other may not be needed. Often, these reaction products have dramatically different physical properties and are easily separable.

$$A + B \longrightarrow C + D$$

Fig. 7.14. Representation of a metathesis reaction.

In many reactions, it is also necessary to anticipate and deal with the products of side reactions. Often, the thermodynamic reaction profile of a synthetic transformation provides alternative mechanistic pathways. These side reactions may generate products of significantly different physical properties, or may give rise to products with physical properties very similar to the desired products. In these cases, complicated purification processes must be developed to isolate the desired compound.

From an economical as well as an environmental standpoint, complicated purification processes should be kept to a minimum. There are many techniques for purification, including recrystallization, precipitation, distillation, fractionation, and chromatography. In each of these techniques the labor costs alone can become quite high. When one considers the additional use of solvents, reagents, energy, and equipment it becomes obvious that such consequences should be addressed early on in the synthetic design. In the review of lengthy synthetic pathways to target molecules, it can be observed that some pathways utilize several iterations of oxidations and reductions, moving repeatedly back and forth between oxidation states. This inefficient process should be minimized when possible and the oxidation state maintained, when feasible, to avoid unecessary redox chemistry.

8 *Evaluation of methods to design safer chemicals*

The methods to design safer chemicals are based on an analysis of how the molecular structure works in achieving its function, versus its ability to cause harm to human health and the environment. It is through the manipulation of the structure, such that the efficacy of function is maximized while the intrinsic hazard of a substance is minimized, that the design of a safer chemical is achieved. There is a common saying among designers of everything from automobiles to furniture to appliances; 'form follows function'. This statement is as true at the molecular level as it is at the macro level.

The question therefore becomes, what is the function one wants to accomplish? This was a straightforward question prior to the time when society became aware of the unforeseen consequences of some chemical products. If one wanted to create a certain shade of red dye, then a molecular structure could be and was created to produce that red dye. If one wanted to create an effective pesticide, then a molecular structure could be generated to serve that function, and so on. The problem as we know it now is not that the solution was incorrect, it was that the question was incorrect and/or incomplete. The true question is therefore 'How do I make a red dye that does not cause cancer?' and 'How do I create an effective pesticide that does not cause environmental damage to birds and other wildlife?'. It is through this expanded consideration and definition of the function that you are trying to accomplish that designing safer chemicals is achieved. As part of the design of any chemical product, the provision that it does not harm human health and the environment must be included.

Once the desired function is decided upon and correlated to a molecular structure, one can then make efforts to adjust and modify the molecular structure to mitigate any potential toxicity or other hazard. There are several basic ways that this can be achieved:

1. mechanism of action analysis
2. structure–activity relationships
3. avoidance of toxic functional groups
4. minimizing bioavailability
5. minimizing auxiliary substances

These different techniques can be implemented depending on several factors, including:

1. How much information you have about the way in which the substance acts as a toxicant?
2. What parameters are known about the substance, such as physical/chemical properties?
3. What information exists on related compounds of the same chemical class or analogous chemical classes?

The more that is known about the specifics of how a chemical substance exhibits its toxicity, the more options are available in designing a safer chemical. As this information becomes less detailed and less specific, the more the molecular designer is forced into knowing what 'not to do' rather than knowing what to do in order to ensure performance with minimal hazard.

8.1 Mechanism of action analysis

While there are chemical substances that are totally inert when in biological systems such as the body or the environment, the vast majority of chemical compounds exhibit some type of biological activity. Certainly, these include all of the food, nutrients, vitamins, and the plethora of other beneficial chemicals that make life itself possible. There is also a subset of chemicals that are of concern for the toxic effects they can cause when introduced into the body or an ecosystem. Each substance has a mechanism by which it produces this undesirable end-point and the more one is able to understand this mechanism, the more one knows how to design the chemical so that this mechanism is either averted or minimized.

A substance can cause the resulting end-point either through direct or indirect toxicity. In the case of direct toxicity, it is the chemical substance itself that is reacting to cause the end effect of concern, whereas with indirect toxicity, it is a metabolite or derivative of the original substance that is responsible for the harmful interaction with the body. It is as a result of recent advances in pharmacology, where the specific steps leading to a toxic end-point have been elucidated, that the reaction path to toxicity has become manifest and can be averted.

Once the mechanism is elucidated, the chemist designing the molecule has several options available to make the chemical intrinsically less hazardous. The structure of the molecule can be changed such that the mechanism of action is no longer possible, as is illustrated by the following example.

It is known that nitrile results in toxicity in a biological system because of the release of cyanide into the body, at which point the well-documented, acute toxicity of cyanide follows (Fig. 8.1). It has been shown that the mechanism of action that a nitrile follows includes the initial formation of a radical in the position α to the cyano group. Following the formation of that radical, the cyanide is then cleaved from the molecule and the toxic end-point results (Fig. 8.2). If one then blocks the α position from forming a radical with the addition of substitutents, e.g. methyl groups, then the mechanism of toxicity is unable to be pursued and the substance is virtually non-toxic (Fig. 8.3). This is borne out by the data on nitriles, which demonstrate a direct correlation between the toxicity of a nitrile and its ability to form and stabilize the α radical.

$$R{-}CH_2{-}CN \longrightarrow \text{"}R{-}CH_2\text{"} + \text{"}CN\text{"}$$

Fig. 8.1. Release of cyanide from a nitrile.

$$R{-}\dot{C}H{-}CN$$

Fig. 8.2. Formation of a radical from a nitrile.

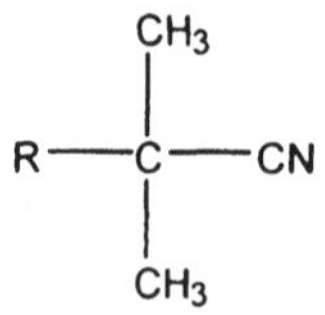

$$R-\overset{\displaystyle CH_3}{\underset{\displaystyle CH_3}{\overset{|}{\underset{|}{C}}}}-CN$$

Fig. 8.3. Blocking the formation of the free radical using substituents prevents the release of cyanide.

This example illustrates the capability to mitigate toxicity through structural changes that curtail the mechanism of toxic action without sacrificing the efficacy of function of the molecule.

8.2 Structure activity relationships

Structure activity relationships (SAR) are also based on a correlation between the molecular architecture of a compound and its activity. For the purpose of this discussion we focus on the 'activity' in the body or in the environment. With SAR, one can observe how subtle changes in the molecular structure can result in changes in the potency or even the presence or absence of a toxic effect within a chemical class. This correlation can be made even in cases where the mechanism of action of a molecule is not known. Therefore, even though one doesn't have a clear picture of 'why' a correlation exists between toxicity and structural modifications, the fact that they do exist is sufficient to assist the chemist in designing a molecule that minimizes the hazard associated with it.

8.3 Avoidance of toxic functional groups

It is often the case where neither the mechanism of action nor a reliable structure activity relationship exists. In this case, it is still possible to identify the functional group that is causing the toxic effect and avoid the use of that functional group. It is, of course, necessary to match the identification of an alternative functionality with the application that the molecule is to perform.

One example of this type of functional group replacement has been demonstrated in the manufacture of adhesives for applications such as car windshields. Many of these adhesives have been made from isocyanates, which cross-link to form adhesive polyurethanes. There is, however, some concern about the toxicological profile of isocyantes and so there has been research and development activity on how to avoid the use of this functional group. One group has developed an adhesive that is based on acetoacetate esters which will serve as adhesive cross-linking agents but avoid the use of the functional group of concern, i.e. isocyanates.

Another way of dealing with a functional group that possesses some toxicity is to mask the functional group. Masking is the technique of temporarily transforming a functionality for a particular purpose only to recover the original functional group when it is required. By using this strategy, a molecule can be rendered innocuous for the time that people or the environment may be exposed to the substance and the reactive functional group will only be regenerated when it is safely contained.

An example of this methodology is used commonly in the dye industry for molecules containing a vinyl sulfone group. Vinyl sulfones are placed as part of a reactive dye in order to allow the dye to bind covalently to fabric so that the dye will not wash off. When toxicological data made it evident that there were concerns for the health implications of the vinyl sulfone group, dye manufacturers masked the vinyl sulfone as the vinyl sulfone sulfate. The sulfate has drastically lower concerns because it can be generated *in situ* without exposure to humans or the environment.

Since this process engages in making a chemical derivative that is designed only to be destroyed and generates waste in the process, this technique would not rate highly in a formal green chemistry analysis. However, if the dominant concern with regard to the handling and use of a substance is the toxicity of an essential functional group, then masking may be the best alternative available.

8.4 Minimizing bioavailability

No matter how inherently toxic a substance is, it must be able to enter the body in order for it to do damage. This ability to enter the various biological systems and organs is called bioavailability. In the cases where there is a general lack of information on how a substance is causing toxicity, or even what functionality within the molecule is responsible, making the molecule less bioavailable is a technique that can often be utilized to minimize the hazard. Much is known about how molecules enter the body through various routes, e.g. respiratory, dermal, or membrane transport. By using this knowledge, chemists can design molecules so that entry into the biological system is either impaired or eliminated.

As an example, if a polymer is of concern when it enters the body's respiratory system, then a useful piece of information would be that particle sizes that are considered respirable are of a size of 10 μm or less. That being the case, a polymer chemist could design the parameters of the polymer particles such that they would be large enough (i.e. > 10 μm) not to be respirable. Since the substance then lacks bioavailability, it could not cause the toxic end-point of concern.

Other substances that enter the body through the skin have a similar restriction to transport. For dermal entry, compounds often need to possess a certain solubility profile as measured by Log *P*. (Log *P* is the measure of how a substance partitions itself in solution between an aqueous and a lipophilic layer.) By manipulating this physical/chemical property, a chemist can make the substance less able to transport through the dermal membrane and therefore less bioavailable. Again, this substance would be unable to exhibit its inherent toxicity because it could not enter the biological system.

8.5 Minimizing auxiliary substances

It can certainly be the case that a substance may possess little or no inherent toxicity but that it requires the use of associated hazardous substances to carry out its function. As an example, if an innocuous substance needs to be dissolved in a hazardous solvent in order to be used properly, then there is toxicity, albeit indirect, which results from the chemical product. This has been the situation with a variety of paints and coatings over the years which require the use of organic solvents to perform their function. These volatile organic compounds are of concern because of their ability to contribute to air pollution, such as atmospheric ozone. In response to these concerns, chemists have been designing new coatings that have the same properties but can be used in aqueous systems or other matrices that do not use volatile organic solvents.

9 *Examples of green chemistry*

9.1 Examples of green starting materials

There have been significant accomplishments in the area of using environmentally benign feedstocks for the manufacture of a wide variety of chemical products. Many products have traditionally been made from feedstocks that contain characteristics of concern, including toxicity, are depleting of natural resources, and/or are generated via methods that can result in environmental damage. With advances in biotechnology, biocatalysis, and biosynthesis, the use of biologically based feedstocks has been demonstrated as a technologically viable alternative to petroleum feedstocks for a number of chemical processes.

9.1.1 Polysaccharide polymers

Polymers are a very important class of compounds that have broad applications and a wide array of properties that can be exploited. The starting materials for polymers, their monomeric precursors, can run the full range of hazards like any other discreet chemical substances. One approach to using more environmentally benign starting materials for polymers is through the use of polysaccharides as the feedstock. Polysaccharide feedstocks have several advantages which make them appealing from a green chemistry perspective.

Polysaccharides are biological feedstocks and, as such, have the advantage of being renewable, as opposed to those feedstocks that are derived from petroleum and other fossil fuels. There is no toxicological data that would implicate polysaccharides as possessing significant hazards in terms of acute or chronic toxicity to human health and the environment. Accident potential from the use of polysaccharides should be negligible. The work of Gross *et al.* utilizes biosynthetic methods to make the polysaccharide-based polymers, a technique that can often be employed as a substitute for carrying out the same transformations using more hazardous substances.[13] One of the additional environmental advantages of using polysaccharides as feedstocks is the fact that they are biodegradable in the ecosystem after their useful life has passed. This is a distinct advantage over many classes of polymers, which tend to persist in the environment.

9.1.2 Commodity chemicals from glucose

Glucose may be an excellent alternative feedstock for commodity chemicals. Frost has demonstrated a variety of syntheses using glucose as a starting material.[14] Using biotechnological techniques to manipulate the shikimic acid pathway (responsible for making many of the aromatic compounds in nature), compounds such as hydroquinone, catechol, and adipic acid, all of which are important, large-volume chemicals, can be synthesized. The traditional starting material (Fig. 9.1) for these substances is benzene, a known carcinogen. By using glucose in place of benzene (Fig. 9.2), this biosynthetic pathway can help to minimize the use of certain reagents with significant toxicity. This synthesis is also conducted in water instead of organic solvents.

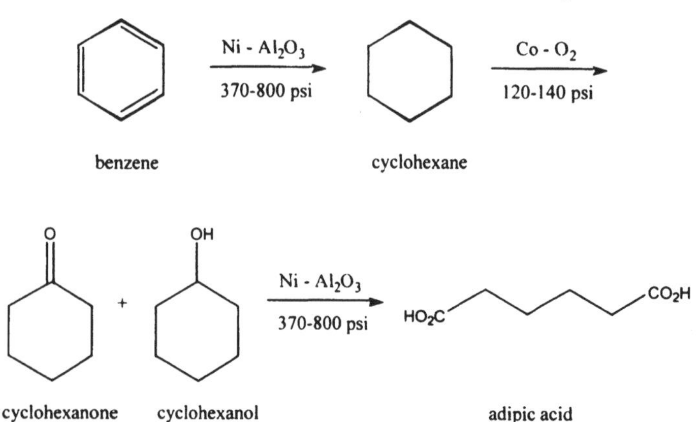

Fig. 9.1. Traditional synthesis of adipic acid, using benzene.

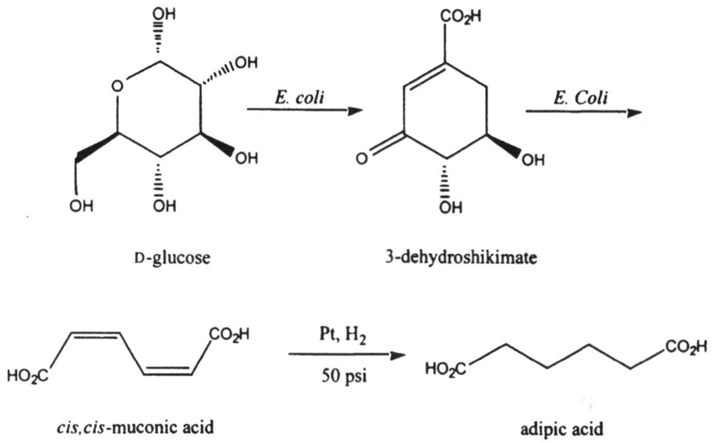

D-glucose → (*E. coli*) → 3-dehydroshikimate → (*E. Coli*) →

cis,cis-muconic acid → (Pt, H_2, 50 psi) → adipic acid

Fig. 9.2. Alternative biosynthetic pathway to adipic acid, using glucose.

9.1.3 Biomass conversion to chemical products

Researchers at Texas A&M University have developed a class of technologies that convert waste biomass into animal feed, industrial chemicals, and fuels (Fig. 9.3).[15] The waste biomass includes such resources as municipal solid waste, sewage sludge, manure, and agricultural residues. These under-utilized resources currently have substantial costs associated with their disposal.

In order to render it more digestible, the waste biomass is treated with lime. Lime-treated agricultural residues such as straw, stover, and bagasse may be used as ruminant animal feeds. Alternatively, the lime-treated biomass can be converted into various chemical products by being fed to a large anaerobic fermentor in which rumen micro-organisms convert the biomass into volatile fatty acid (VFA) salts such as calcium acetate, propionate, and butyrate. The salts are concentrated and may be converted into chemicals or fuels via three routes. In the first route, the VFA salts are acidified, releasing acetic, propionic, and butyric acids. In the second route, the VFA salts are thermally converted to ketones such as acetone, methyl ethyl ketone, and diethyl ketone. In the third route, the ketones may be hydrogenated to their corresponding alcohols such as isopropanol, isobutanol, and isopentanol.

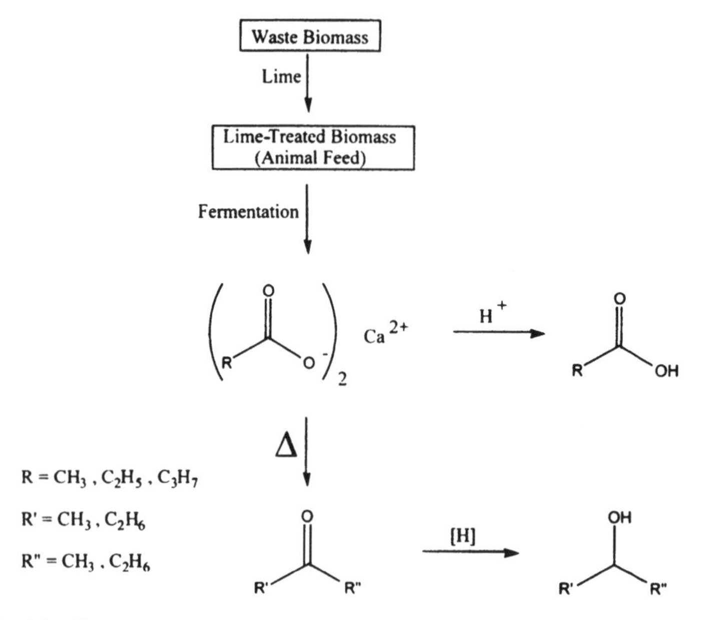

Fig. 9.3. Technologies to convert waste biomass into animal feed, industrial chemicals, and fuels.

This class of technologies offers significant benefits for human health and the environment. Lime-treated animal feed can be used to replace feed corn, which uses approximately 88% of all corn production. The growing of corn requires plowing, which exacerbates soil erosion, and approximately two bushels of top soil are lost for each bushel of corn harvested. Corn also requires intensive inputs of fertilizers, herbicides, and pesticides, which may contribute to contamination of groundwater.

Chemicals may be produced economically from waste biomass that has a negative impact on the environment, such as municipal solid waste and sewage sludge. Typically, these wastes are landfilled or incinerated, which incurs a disposal cost while contributing to land or air pollution. By producing chemicals from biomass, nonrenewable resources such as petroleum and natural gas, are conserved for later generations.

Fuels produced from waste biomass have the benefits cited above, i.e. reduced environmental impact from waste disposal and reduced trade deficit. In addition, oxygenated fuels derived from biomass are cleaner burning and do not add net carbon dioxide to the environment, thereby reducing the factors that contribute to global warming.

9.2 Examples of green reactions

9.2.1 Atom economy and homogeneous catalysis

In order to maximize the number of atoms of reactants that are transformed to products in a synthetic scheme, Trost has developed what is called 'atom economy'.[9] The goal of this work is to reduce the number of atoms that are produced as unwanted by-products. Diels–Alder reactions and aldol condensations are examples where little or no by-products are formed. To improve these and other types of reactions, Trost has been developing elegant transition metal catalysts. This work has been well documented in a recent review.

9.2.2 Halide-free syntheses of aromatic amines

Traditional syntheses of aromatic amines involve chlorination of benzene followed by nitration and nucleophilic displacement of the chlorine with a new substituting group. The synthesis of 4-aminodiphenylamine illustrates this process (Fig. 9.4).

Monsanto has developed a new synthesis of 4-aminodiphenylamine that utilizes nucleophilic substitution for hydrogen (Fig. 9.5).[16] This process avoids the use of halogenated intermediates. In this process nitrobenzene and aniline are heated in the presence of tetramethyl ammonium hydroxide to give the tetramethyl ammonium salts of the condensation products. Catalytic hydrogenation of this reaction mixture affords 4-aminodiphenylamine while regenerating the tetramethyl ammonium hydroxide.

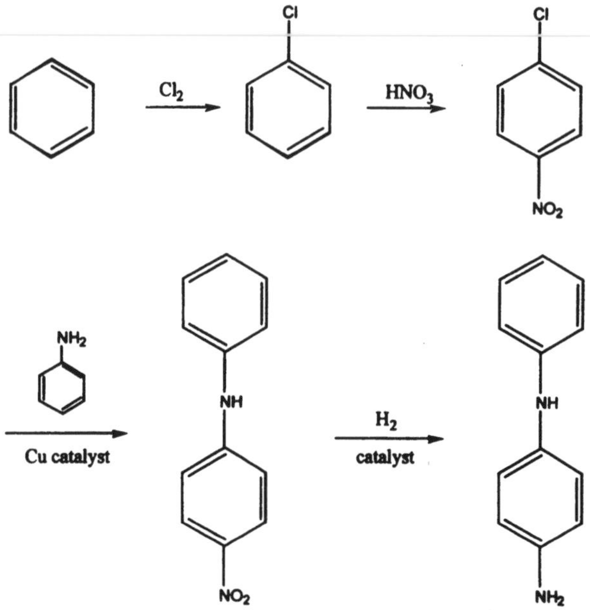

Fig. 9.4. Traditional synthesis of an aromatic amine, 4-aminodiphenylamine.

9.2.3 A green alternative to the Strecker synthesis

Disodium iminodiacetate (DSIDA) is a key intermediate in the production of Monsanto's Roundup(r) herbicide.[17] Traditionally, Monsanto and others have manufactured DSIDA using the well-known Strecker process (requiring ammonia, formaldehyde, hydrogen cyanide, and hydrochloric acid) (Fig. 9.6). Because of its extreme, acute toxicity, hydrogen cyanide requires special handling to minimize the risk to workers, the community, and the environment. The reaction can generate, exothermically, potentially unstable intermediates. The overall process also generates up to 1 kg of waste for every 7 kg of product. Much of this waste contains traces of cyanide and formaldehyde and must be treated prior to safe disposal.

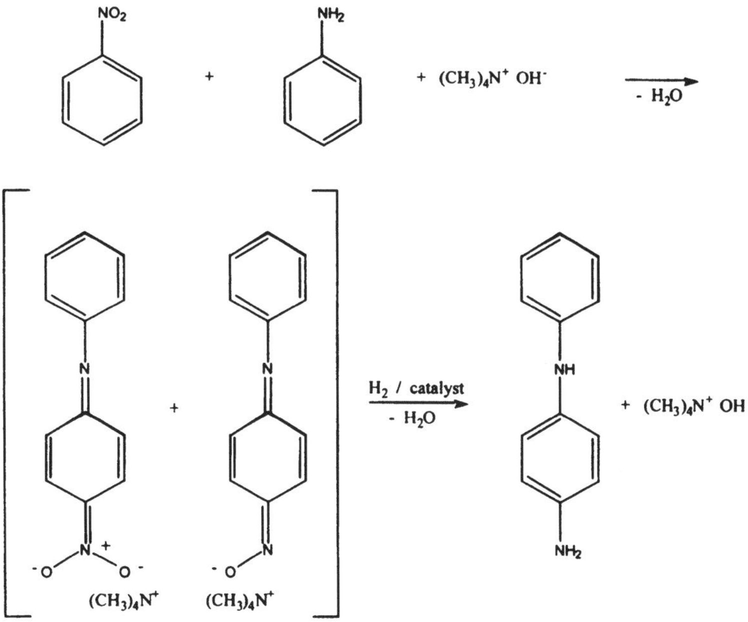

Fig. 9.5. New synthesis of 4-aminodiphenylamine avoiding the use of halogenated intermediates.

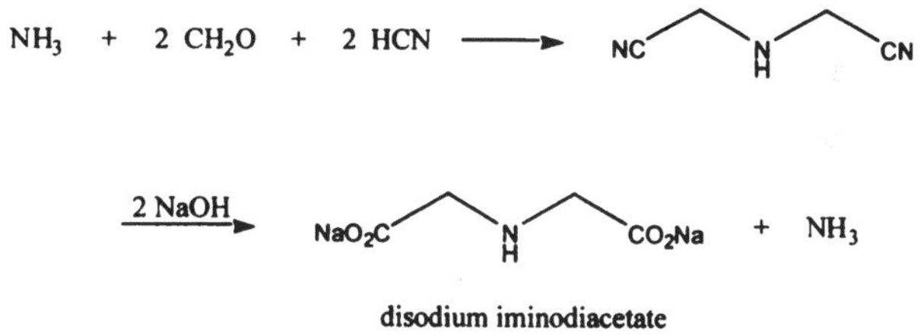

Fig. 9.6. The Strecker process for synthesizing DSIDA.

Monsanto has developed and implemented an alternative DSIDA process that relies on the copper-catalyzed dehydrogenation of diethanolamine (DEA) (Fig. 9.7). This process is inherently safer because the dehydrogenation reaction is endothermic and so does not present the danger of a runaway reaction. This new technology avoids the use of cyanide and formaldehyde, is safer to operate, produces a higher overall yield, has fewer process steps, and produces a product stream that, after filtration of catalyst, is of such high quality that no purification or waste cut is necessary for subsequent use in the manufacture of Roundup(r). This catalysis technology can also be used in the production of other amino acids, such as glycine. It is also a general method for conversion of primary alcohols to carboxylic acid salts, and is potentially applicable to the preparation of many other agricultural, commodity, specialty, and pharmaceutical chemicals.

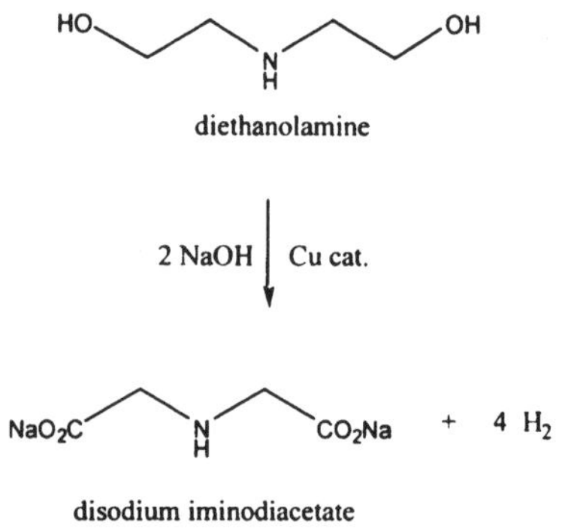

HO–CH₂CH₂–NH–CH₂CH₂–OH

diethanolamine

2 NaOH | Cu cat.

NaO_2C–CH₂–NH–CH₂–CO_2Na + 4 H_2

disodium iminodiacetate

Fig. 9.7. Alternative synthesis of DSIDA using a copper catalyst.

9.3 Examples of green reagents

9.3.1 Non-phosgene isocyanate synthesis

Polyurethanes are important polymers that are widely used for a variety of commercial applications. The environmental difficulty with polyurethanes is that they have been traditionally manufactured through the use of phosgene (Fig. 9.8). Phosgene is an extremely toxic gas whose acute end-point is lethality. A group at the Monsanto Company has developed a method of synthesizing polyurethanes and their isocyanate precursors that totally eliminates the use of phosgene (Fig. 9.9).[18]

$$\underset{\text{amine}}{RNH_2} + \underset{\text{phosgene}}{COCl_2} \longrightarrow \underset{\text{isocyanate}}{RNCO} + 2\,HCl \xrightarrow{R'OH} \underset{\text{urethane}}{RNHCO_2R}$$

Fig. 9.8. Synthesis of polyurethanes using phosgene.

$$\underset{\text{amine}}{RNH_2} + \underset{\text{carbon dioxide}}{CO_2} \longrightarrow \underset{\text{isocyanate}}{RNCO} + H_2O \xrightarrow{R'OH} \underset{\text{urethane}}{RNHCO_2R'}$$

Fig. 9.9. Alternative synthesis of polyurethanes without using phosgene.

9.3.2 Selective methylations using dimethylcarbonate

Conventional methylation reactions employ methyl halides or methyl sulfate. The toxicity of these compounds and their environmental consequences render these syntheses somewhat undesirable. The methylation of active methylene compounds often involves uncontrollable multiple alkylations.

Tundo has developed a method to methylate active methylene compounds selectively using dimethylcarbonate (Fig. 9.10).[19] Reacting arylacetonitriles with dimethylcarbonate at 180–220 C in the presence of potassium carbonate produces 2-arylpropionitriles with high selectivity (>99%). In this process no inorganic salts are produced. Tundo has demonstrated this reaction under both continuous-flow and batchwise conditions.

CN
R
+
O
H_3C O O CH_3
K_2CO_3
H_3C CN
R
+ CH_3OH + CO_2

Fig. 9.10. Selective methylation of active methylene compounds.

9.3.3 Solid-state polymerization of amorphous polymers using diphenylcarbonate

The Asahi Chemical Industry has developed the concept of solid-state polymerization of amorphous polymers into the manufacture of polycarbonates.[20] Replacing the conventional use of phosgene and methylene chloride, this process uses bisphenol-A and diphenyl carbonate directly to give low molecular weight 'prepolymers' (MW 2000–20 000) (Fig. 9.11). These 'prepolymers' are then converted to higher molecular weight, optically transparent polymers by crystallizing the low molecular weight material, followed by further polymerization.

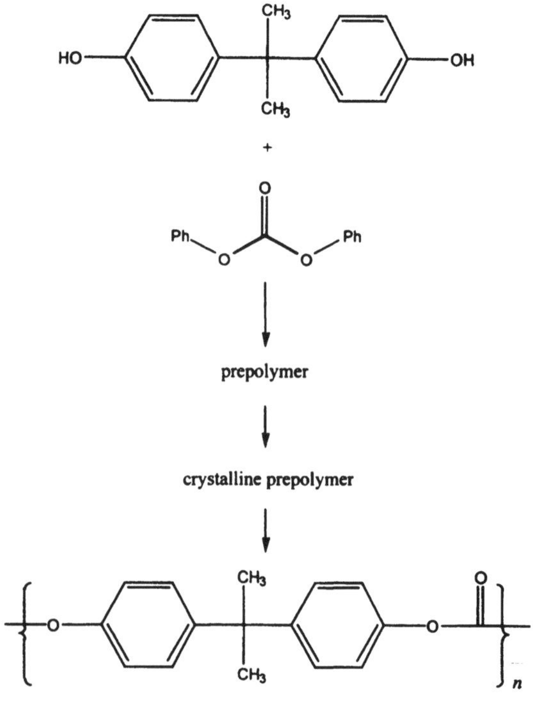

HO–C₆H₄–C(CH₃)₂–C₆H₄–OH

+

PhO–C(=O)–OPh

↓

prepolymer

↓

crystalline prepolymer

↓

$[-O-C_6H_4-C(CH_3)_2-C_6H_4-O-C(=O)-]_n$

Fig. 9.11. Synthesis of low molecular weight 'prepolymers'.

9.3.4 Green oxidative transition metal complexes

Many oxidative processes have negative environmental consequences. By creating long-lived catalytic and recyclable oxidants, metal ion contamination in the environment can be minimized by using molecular oxygen as the primary oxidant. Several ligand systems that are stable towards oxidative decomposition in oxidizing environments are being developed by Collins.[21] Modeling proposed reactive intermediates in biological and chemical oxidation processes, extremely stable high oxidation state transition metal complexes have been synthesized.

9.3.5 Liquid oxidation reactor

Praxair has developed a process that allows the safe oxidation of organic chemicals with pure or nearly pure oxygen.[22] This technology, known as the liquid oxidation reactor (LOR), provides environmental advantages. The use of oxygen in place of conventional, air-based oxidation processes reduces the amount of vent gas that must be treated prior to atmospheric release. The use of oxygen can have a positive effect on the chemistry of the reaction by allowing lower temperatures and/or pressures. This can lead to an improvement in selectivity without sacrificing production rate. The increased chemical efficiency with oxygen can result in substantial raw materials cost savings. The lower temperature allowed by the LOR process reduces the loss of reactant and/or solvent to by-products and to waste streams. The LOR will enable a large and important segment of the US chemical industry to realize more efficient use of raw materials, reduced environmental emissions, and energy savings.

9.4 Examples of green solvents and reaction conditions

There has been a large emphasis, both in the chemical industry and academic research, on the development of environmentally benign solvents and reaction conditions. This is largely owing to the fact that traditional solvents, such as chloroflurocarbons and volatile organic compounds, have been implicated in a number of environmental problems and so have been highly regulated. The accomplishments in this area have been innovative and diverse and responsive to the economic and legislative demands imposed on the use of the classical solvent systems.

9.4.1 Supercritical fluids

9.4.1.1 Asymmetric catalysis using supercritical carbon dioxide

The use of supercritical carbon dioxide as a substitute for organic solvents already represents an important tool for waste reduction in the chemical industry and related areas. Coffee decaffeination, hops extraction, and essential oil production, as well as waste extraction/recycling and a number of analytical procedures, already use this non-toxic, non-flammable, renewable, and inexpensive compound as a solvent. The extension of this approach to chemical production, using CO2 as a reaction medium, is a promising approach to pollution prevention. Of the wide range of supercritical carbon dioxide reactions that have been explored, one class of reactions has shown exceptional promise.[23] Los Alamos National Laboratory has found that asymmetric catalytic reductions, particularly hydrogenations and hydrogen transfer reactions, can be carried out in supercritical carbon dioxide with selectivities comparable or superior to those observed in conventional organic solvents (Fig. 9.12).

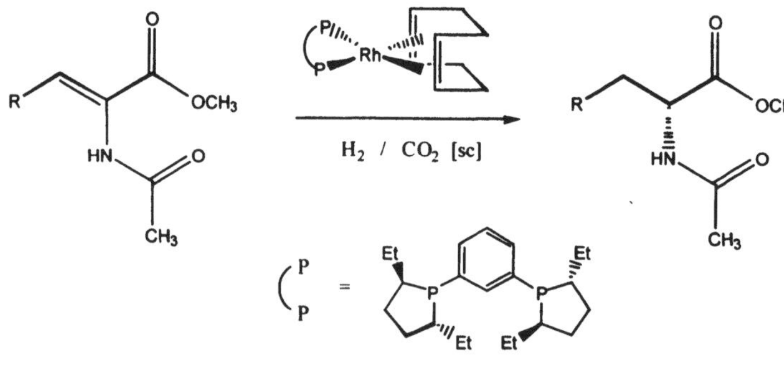

Fig. 9.12. Asymmetric catalytic reduction in supercritical carbon dioxide.

Los Alamos has discovered, for example, that asymmetric hydrogen transfer reduction of enamides using ruthenium catalysts proceeds with enantioselectivities that exceed those in conventional solvents. The success of asymmetric catalytic reductions in CO2 is due in part to several unique properties of CO2, including tunable solvent strength, gas miscibility, high diffusivity, and ease of separation. In addition, the insolubility of salts, a significant limitation of CO2 as a reaction solvent, has been overcome by using lipophilic anions, particularly tetrakis[3,5-bis(trifluoromethyl)-phenyl]borate (BARF). These discoveries demonstrate an environmentally benign and potentially economically viable alternative for the synthesis of a wide range of specialty chemicals such as pharmaceuticals and agrochemicals.

9.4.1.2 Supercritical polymerizations

Carbon dioxide surfactant technology, or 'soapy CO2', uses liquid/supercritical CO2 in place of less acceptable organic chemicals.[24] Carbon dioxide represents an environmentally friendly alternative to the solvents currently used in a variety of applications. This technology involves the development of surfactant systems for CO2 in order to expand the use of liquid and supercritical CO2 to enhance its solvating power for large, hydrocarbon-based molecules. In addition to polymerization processes (Fig. 9.13), 'soapy CO2' can be used as a cleaning and extraction medium (replacing halogenated hydrocarbons) as well as a solvent/medium for organic reactions.

Fig. 9.13. Use of 'soapy CO2' in polymerization.

9.4.1.3 Free radical brominations in supercritical CO2

Another example of replacing conventional solvents with supercritical CO2 has been described by Tanko.[25] His work has demonstrated that free radical brominations can be performed in supercritical CO2, where selectivity and yield are not compromised by switching from conventional reaction conditions.

The free radical bromination of toluene, for example, was explored using both bromine and *N*-bromosuccinimide (NBS) (Fig. 9.14). When bromine was used as the brominating agent, a mixture of benzyl bromide (>70%) and 4-bromotoluene was obtained. When *N*-bromosuccinimide was used, quantitative conversion to benzyl bromide was observed.

Fig. 9.14. The free radical bromination of toluene in supercritical carbon dioxide.

9.4.1.4 Carbon dioxide blowing agents

Becuase of environmental concerns, the Dow Chemical Company has developed a novel process for the use of 100% carbon dioxide CO2 as a blowing agent used to manufacture polystyrene foam sheet.[26] The use of 100% CO2 offers optimal environmental performance because CO2 does not deplete the ozone layer, does not contribute to ground-level smog, and will not contribute to global warming since the CO2 is used from existing by-product commercial and natural sources. Carbon dioxide is also non-flammable, providing increased worker safety, and is cost effective.

9.4.2 Aqueous reaction conditions

9.4.2.1 Breslow's cyclodextrin work

There have been a number of investigations of using water as a solvent in place of traditional organic solvents as the medium for carrying out synthetic transformations. Breslow has researched the

Diels–Alder reaction in water and found that the rates of reaction and selectivity can be increased in an aqueous system.[27] Breslow has also investigated the use of cyclodextrin to facilitate the use of water as a reaction medium and has carried out transformations such as catalytic cytochrome P-450 biomimetic reactions.

9.4.2.2 Aqueous-based indium catalysis

The metal indium, a relatively unexplored element, has recently been shown by Paquette to offer advantages for promoting organic transformations in aqueous solution.[28] The feasibility of performing organometallic/carbonyl condensations in water, for example, has been demonstrated for the metal indium. Indium is non-toxic, very resistant to air oxidation, and easily recovered by simple electrochemical means, thus permitting its re-use and guaranteeing uncontaminated waste flow. In addition, protection/deprotection of functional groups and an inert atmosphere are not necessary when implementing this technology.

9.4.2.3 Aqueous free radical bromination

Multiphase reactors have been developed at the New Jersey Institute of Technology and other universities that use water as the reaction medium in order to avoid the use of hazardous organic solvents in the manufacture of pharmaceuticals and specialty chemicals.[29] This technology demonstrates that free radical bromination of organics can be carried out in aqueous systems. A unique semi-continuous droplet reactor has also been developed for epoxidations. That these methods offer a new 'non-end-of-pipe' method of eliminating VOC's is a major incentive for process modification.

9.4.3 Immobilized solvents

A class of polymeric solvents has been developed by researchers at MIT that have solvation properties similar to those of the solvents used conventionally in chemical synthesis, separations, and cleaning operations.[30] The potential for loss by environmentally unfavorable air emissions or aqueous discharge streams is minimized. These solvents are polymeric derivatives of solvents currently used in reaction and separation processes. The solvents can be used as neat reaction or separation media, or they can be diluted in higher alkanes. Polymeric or oligomeric solvents such as derivatized THF (tetrahydrofuran) have been synthesized using macromonomers incorporating the desired functionality as pendant groups on the polymeric backbone (Fig. 9.15). These polymeric solvents are easily recovered using mechanical separations such as ultrafiltration rather than distillation processes. This concept for solvent design and synthesis offers the potential for significant source reductions in air and water pollution.

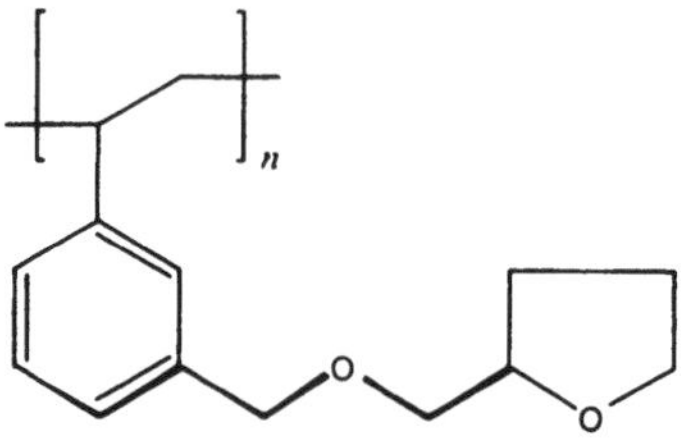

Fig. 9.15. A macromonomer used to synthesize polymeric or oligomeric solvent.

9.4.4 Irradiative reaction conditions

9.4.4.1 Photosensitized cleavage of the dithio protecting group by visible light

The design of environmentally benign photochemical reactions through the use of non-toxic alternatives for the oxidation of dithianes, oxathianes, and benzyl ethers has been described by Epling.[31] In one example dithio derivatives of carbonyl compounds are deprotected using visible light (Fig. 9.16). By using a conventional 120 Watt spotlight and methylene green as a sensitizer, dithio derivatives have been converted to their corresponding aldehydes and ketones in excellent yields.

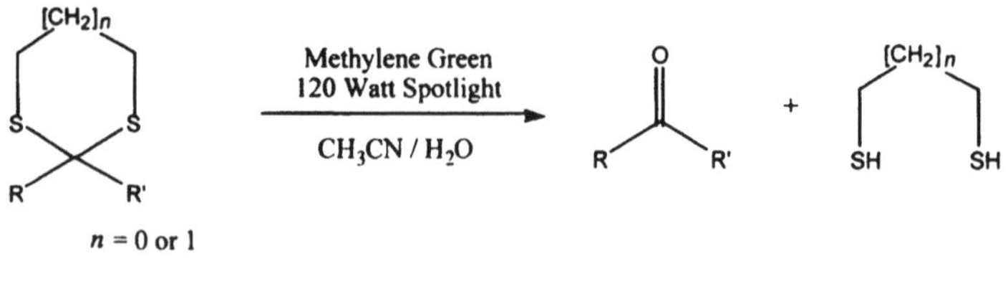

Fig. 9.16. Deprotection of dithio derivatives of carbonyl compounds using visible light.

9.4.4.2 Photochemical alternative to the Friedel–Crafts reaction

In order to avoid the generation of the pollutant by-products of conventional Friedel–Crafts reactions, an alternative method has been described by Kraus.[32] By the photochemically mediated reactions of aldehydes with quinone, benzodiazepine, and benzoepin, ring systems have been synthesized (Fig. 9.17). This methodology avoids the use of air-sensitive acid chlorides, Lewis acid catalysts (aluminum chloride, stannous chloride, or titanium chloride), or solvents such as nitrobenzene, carbon disulfide, or carbon tetrachloride.

9.5 Examples of green chemical products

Since the design of safer chemicals can apply to virtually any chemical product, the examples given here will be as diverse as the products themselves. While it is true that nothing is risk-free, the advantages offered by employing the tools for designing safer chemicals allows for drastically reduced hazards and for the products synthesized to be environmentally benign.

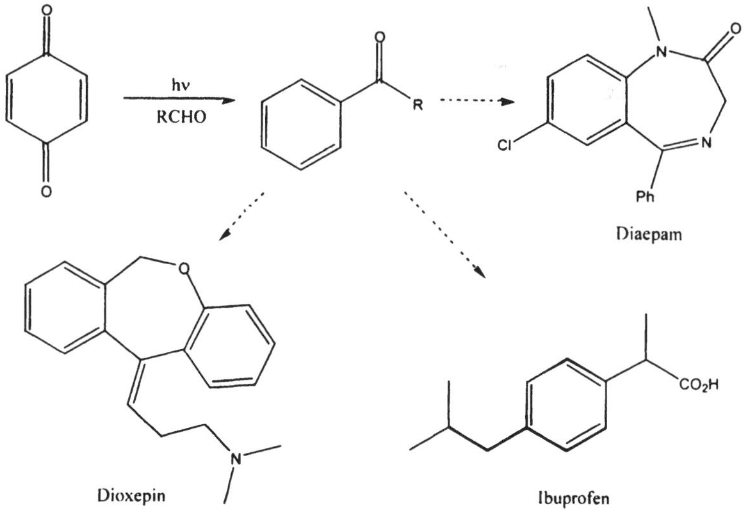

Fig. 9.17. The photochemically mediated reaction of aldehydes with quinone.

9.5.1 Design of alternative nitriles

An example of how one can assess the toxicological consequences of a particular functional group and then use this information to design inherently safer compounds has been described by DeVito, who presents a comprehensive study of the nitrile functional group.[33] Toxicological structure–activity relationships are explored and synthetic modifications that reduce toxicity are found. The general strategy described can be applied to the study of any class of compounds.

One of many examples presented in DeVito's work is a comparison of the toxicity of 3-hydroxypropionitrile vs. 2-hydroxypropionitrile (Fig. 9.18). Toxicological studies demonstrate that the 2-hydroxy isomer is much more toxic than the 3-hydroxy isomer. The mechanism of acute toxicity is proposed to be elimination of hydrogen cyanide from the cyanohydrin (Fig. 9.19). Depending on the nature of the substitution at the α carbon position, this elimination can be accelerated or slowed down. Knowing this, new propionitriles can be designed to reduce this mechanistic pathway and thus be inherently safer.

CH_2—CH_2—CN (with OH on the first CH_2) H_3C—CH—CN (with OH on CH)

3-Hydroxypropionitrile 2-Hydroxypropionitrile

rat oral LD_{50} = 45 mmol/kg rat oral LD_{50} = 1.23 mmol/kg

Fig. 9.18. The toxicity of 3- and 2-hydroxypropionitrile.

R'—C(R)(OH)—CN → R'C(=O)R + HCN

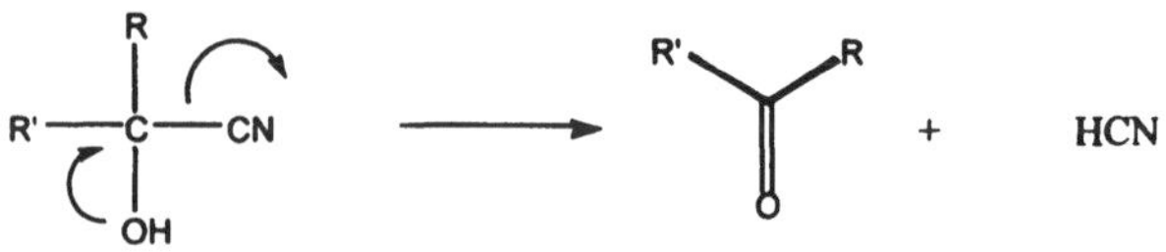

Fig. 9.19. Elimination of hydrogen cyanide from the cyanohydrin.

9.5.2 Rohm and Haas Sea-Nine(tm) product

Fouling is the unwanted growth of plants and animals on a ship's surface. Chemicals used to control fouling are organotin compounds such as tributyltin oxide (TBTO). They are effective at preventing fouling, but have widespread environmental problems. They are persistent in the environment and have several toxic effects, including acute toxicity, bioaccumulation, decreased reproductive viability, and increased shell thickness in shellfish. These harmful effects led to the Organotin Antifoulant Paint Control Act of 1988. This act mandated restrictions on the use of tin in the United States and charged the EPA and the US Navy with conducting research on alternatives to organotins.

The ideal antifoulant would prevent fouling from a wide variety of marine organisms without causing harm to non-target organisms. 4,5-Dichloro-2-*n*-octyl-4-isothiazolin-3-one (Sea-Nine(tm) antifoulant) was chosen by the Rohm and Haas Company as a new antifoulant for commercial development.[34]

Sea-Nine(tm) antifoulant degrades extremely rapidly with a half-life of one day in sea water and one hour in sediment. TBTO degrades much more slowly, with a half-life in sea water of nine days and of six to nine months in sediment. Sea-Nine(tm) antifoulant's bioaccumulation is essentially zero.

9.5.3 Rohm and Haas CONFIRM(tm) insecticide

CONFIRM(tm) effectively and selectively controls important caterpillar pests in agriculture without posing significant risk to the applicator, the consumer, or the ecosystem.[35] This insecticide controls caterpillars through an entirely new and inherently safer mode of action than current insecticides. It acts by mimicking a natural substance found in the caterpillar's body, called 20-hydroxyecdysone, which is the natural 'trigger' that induces molting and regulates development in insects. CONFIRM(tm) disrupts the molting process in caterpillar pests, causing them to stop feeding within hours of exposure and to die soon after. This compound is a poor mimic in most other insects and arthropods and so it is safe to a wide range of beneficial, predatory, and parasitic insects such as honeybees, ladybeetles, parasitic wasps, predatory bugs, and lacewings. CONFIRM(tm) does not bioaccumulate, volatilize, leach, or persist unreasonably long in the environment.

9.5.4 Donlar's polyaspartic acids

Polyacrylic acid (PAC) is an important anionic polymer used in many industrial applications. The ideal disposal for these polymers would be via biodegradation by microorganisms because the degraded end-products are innocuous. Unfortunately, PAC is not biodegradable, so in most cases, these polymers end up in waste treatment facilities. An economically viable, effective, and biodegradable alternative is thermal polyaspartate (TPA).

Donlar has invented two highly efficient processes to manufacture TPA.[36] The first involves a dry and solid polymerization, converting aspartic acid to polysuccinimide. No organic solvents are involved during the conversion. The by-product is condensed water. This process is extremely efficient. Yields of more than 97% of polysuccinimide are achieved routinely. The second step of this process, the base hydrolysis of polysuccinimide to polyaspartate, is also extremely efficient and waste free.

The second TPA production process involves using a catalyst during the polymerization that allows a lower heating temperature to be used. The resulting product has improved performance characteristics, lower color, and biodegradability. Because the catalyst itself can be recovered from the process, waste is further minimized.

TPA is non-toxic and environmentally safe. TPA biodegradability has also been tested using established OECD (Organization for Economic Cooperation and Development) methodology. PAC cannot be classified as biodegradable when tested under the same conditions.

9.5.5 Polaroid's complexed developers

To obtain photographic developers that have the correct physical and chemical properties, Polaroid, along with others in the photographic industry, have covalently modified hydroquinones. Typically, blocking groups are attached to the hydroquinone oxygens (Fig. 9.20).

Taking advantage of the alkali present in the 'pod' reagent, base-reactive functionalized substitutents are used. Acyl hydrolysis and elimination mechanisms have been employed to 'mask' the hydroquinones and render them non-reactive to oxidation. Prior to activation by base release of the protecting group, the substituted hydroquinone is oxidatively inert.

From an environmental perspective this approach is not satisfactory. To arrive at a molecule whose base-release kinetics are precisely matched to the thermodynamic and kinetic requirements imposed by the photographic system, several candidate molecules will typically have to be synthesized, purified, screened, and tested. The toll of this procedure in terms of organic solvent utilization and hazardous waste generation, in addition to the associated economic cost, adds up quite rapidly. In addition, because the diffusion problems are not efficiently addressed, the solubility properties of such molecular systems can require coating solutions to incorporate non-aqueous solvents.

BLPG = Base Labile Protecting Group

Fig. 9.20. Covalently modified hydroquinones.

To address these issues, the use of molecular recognition and self-assembly as a means to control chemical behavior has been applied.[37] The general idea is to form a supramolecular construct whose structural integrity is maintained by non-covalent interactions. Using dialkylterephthalamides as the conjugating component, molecular assemblies can be constructed. The tertiary amide–phenol hydrogen bond provides a base-labile force of attraction enhanced by the stacking available between the electron-rich hydroquinone and the electron-deficient terephthalamide (Fig. 9.21).

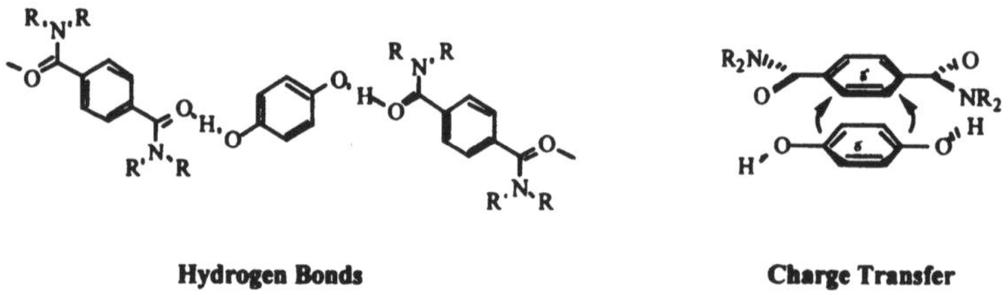

Hydrogen Bonds **Charge Transfer**

Fig. 9.21. Intermolecular interactions

The solid-state behavior of the supramolecular construct is dramatically different from the individual components. While the hydroquinones have significant solubility in water, the complex of hydroquinone and terephthalamide is relatively water insoluble. This insolubility allows for the processing of this material in water. Because the strength of the solid-state complex is maintained by hydrogen bonds, when the alkali from the reagent 'pod' raises the pH of the film, the hydroquinones are deprotonated and rapidly solubilized, and thus activated for use in the photographic system (Fig. 9.22).

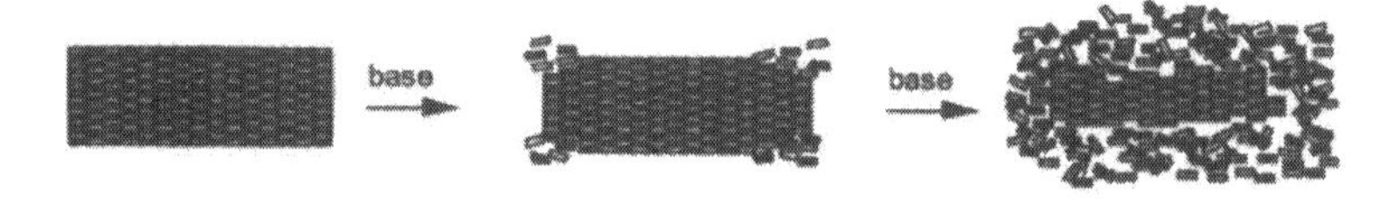

Fig. 9.22. Representation of the deprotonation and solubilization of the hydroquinone and terephthalamide complex.

Analytical measurements have demonstrated that auto-oxidation of the hydroquinones is significantly reduced or eliminated when fully complexed in a supramolecular assembly. Evaluation of hydroquinone concentrations in various film layers after coating has verified that the complexed developers do not migrate throughout the structure and are efficiently immobilized. That the photographic process works as well as it does, clearly shows that the assembly is solubilized and the hydroquinone is released in response to the pH change initiated during the photographic process.

10 *Future trends in green chemistry*

The future of green chemistry is as broad as the future of chemistry as a whole and is therefore difficult to predict or summarize. Just as chemistry has always been a journey rather than a conclusion, green chemistry is also based on the premise that continual improvement, discovery, and innovation is the path towards the perfect goal of environmentally benign.

In this context there are, however, some areas of research that pose both a scientific challenge to chemists and have the potential for dramatic benefits as green chemistry alternatives to the current state of science.

10.1 Oxidation reagents and catalysts

While there have been notable advances in recent years, oxidation chemistry has had the mixed blessing of being both one of the most essential and one of the most polluting chemical technologies. It is certainly essential because oxidative transformations have been the basis of much of the necessary functionalization of fundamental chemical building blocks. Since the chemical industry currently relies almost exclusively on petroleum-based feedstocks, which are of course in their nearly fully reduced state, it is oxidation chemistry that allows these feedstocks to become the chemical products ultimately used by the consumer.

Historically, many of the oxidation reagents and catalysts have been comprised of toxic substances such as heavy metals (e.g. chromium). Since these substances were often used in extremely large volumes required to convert billions of pounds of petrochemicals, there was a significant legacy of these metals being released to the environment and having substantial negative effects on human health and the environment.

In order to change this profile of oxidation chemistry, much attention has been paid to the development of green chemistry techniques that will be both environmentally benign as well as bringing other efficiency, selectivity, and, hence, economic benefits. The new oxidation chemistries will almost certainly need to be catalytic rather than stoichiometric and be very robust, with high turnover rates. If heavy metals/transition metals are used at all, they will certainly be utilizing some of the most innocuous metals, e.g. iron. The basis for the oxidations may often involve molecular oxygen and/or peroxide.

The key to the new green oxidation chemistry will, of course, be the use and generation of little or no hazardous substances, with maximum efficiency of atom incorporation. This is an area of interest, which should yield significant results in the years to come, and which will have a dramatic effect on all types of products, processes, and industrial sectors.

10.2 Biomimetic, multifunctional reagents

Nature has taught us some of its most profound lessons and that is as true in the field of chemistry as it is in every other aspect of life. As scientists elucidate the mechanisms that biological systems use to carry out their functions, a template is created that can be used in designing reagents of the future. This 'biomimetic' approach to the design of catalysts and reagents embraces some common and admirable features found in biological systems such as enzymes.[38]

While synthetic catalysts and reagents for the most part have centered on carrying out one discrete transformation (e.g. reduction, oxidation, methylation), biochemical systems often carry out several manipulations with the same reagent. These manipulations may include activation, conformational adjustments, and one or several actual transformations and derivitizations.

10.3 Combinatorial green chemistry

Combinatorial chemistry is the practice of being able to make large numbers of chemical compounds rapidly on a small scale through reaction matrices. This practice has found widespread adoption, especially in the pharmaceutical sector, and the advantages to green chemistry have been cited. If a pharmaceutical company identifies a 'lead compound', i.e. one with considerable promise, then, historically, the company would proceed in making a large number of derivatives of the lead in order to test their efficacy and optimize the potential. The advent of combinatorial chemistry has enabled large numbers of substances to be made and their properties assessed without the magnitude of the effects of waste and disposal of associated material being as large as it has been the case in the past.

10.4 Chemistry that both prevents problems and solves current pollution problems

It is now being discovered that many of the green chemistry technologies that have been discovered or pursued for pollution prevention can simultaneously be used to deal with environmental problems that already exist. One example of this might be the use of carbon dioxide as an alternative feedstock in the building materials industry. This work utilizes CO2 in a manner that greatly increases the performance characteristics of materials such as concrete or other materials that can be used in wallboard, by the incorporation of carbon dioxide into the material. By finding a use for carbon dioxide that sequesters it in a matrix, this technology is not only pollution preventing, it also deals with the greenhouse gas problem that already exists by reducing the amount of CO2 released to the atmosphere.

Another example of the use of an alternative feedstock that is both green chemistry and remediates an existing environmental problem is that of biomass utilization for the manufacture of chemicals. The biomass problem as a solid waste issue has been around for years. Depending on the type of waste, it can be responsible for filling municipal landfills at an alarming rate. Through the use of technologies that consume biomass in the manufacture of value-added products, this green chemistry alternative for pollution prevention is also a pollution remediation technology.

Other wastes, such as halogenated aromatics, which are by-products of many chemical processes, especially in the pharmaceutical industrial sector, are now finding uses as feedstocks through the use of new biocatalytic technologies. Again, the wastes that exist and are being generated are costly both to treat and to dispose of. By using the new green chemistry technologies, these existing environmental problems can begin to be dealt with.

10.5 Proliferation of solventless reactions

One of the 'solvent alternatives' that is being developed in green chemistry is that of solventless or neat reaction systems. Reactions and whole manufacturing processes are being conducted in solventless conditions such as molten-state reactions, dry-grind reaction, plasma, and neat solid-supported reactions such as clay and zeolites. These techniques are utilizing some non-traditional conditions such as microwave, ultrasound, and visible light transformations. As the development of additional transformations and entire solventless synthetic pathways are developed, this area will need to develop methodologies for product isolation, separation, and purification that will be solventless as well in order to maximize the benefits.

10.6 Energy focus

The environmental effects of energy usage are profound but have not been as visible and as direct as some of the hazards that have been posed by materials used in the manufacture, use, and disposal of chemicals. The new focus on the energy implications of chemical transformations needs to be a key focus of fututre research in green chemistry. The energy benefits of catalysis have been dramatic, especially in the area of petrochemistry. The need to design substances and materials that are effective, efficient, and inexpensive at the capture, storage, and transport of energy is a major challenge for green chemistry. The hazards, whether direct or indirect, to human health and the environment from the utilization of inefficient and polluting energy sources are ones that can and must be adressed by green chemistry in the future.

10.7 Non-covalent derivatization

The manufacture, processing, and use of chemicals has depended largely on a system of making and breaking covalent bonds. Certainly, one could say that synthetic organic chemistry defined itself as the making and breaking of carbon–carbon bonds. This paradigm must change. Chemistry happens without bond making; physical/chemical properties are modified and performance measures are enhanced. Through the utilization of dynamic complexation, which allows for the temporary formation of modified chemical structures, the properties of a molecule can be changed for the period of time necessary to carry out a particular function without all of the waste that would be generated if full derivatization were implemented.[39]

Exercises

Chapter 1

1. Identify one of the major environmental laws. How does it address environmental problems? Through waste treatment? Regulation of emissions? Is the law an example of 'command and control'? Why?
2. Identify an environmental incident or condition that mobilized public opinion to establish a new law.
3. Identify an example of pollution prevention being used in business. What are its environmental and economical advantages?
4. What year was the Pollution Prevention Act passed and what hierarchy of approaches to environmental problems did it establish?
5. What are the traditional factors considered by chemists in the design of chemical syntheses?
6. What are the hazards posed by CFCs?
7. Were the impacts of CFCs known at the time of their discovery?
8. What is diozin and what toxic end-point does it cause?

Chapter 2

1. What are the basic factors involved in calculating risk?
2. How does green chemistry approach the goal of risk reduction?
3. Identify the various types of hazards that a chemical substance can possess.
4. What are the three components of a hazard that need to be assessed?
5. What are the disadvantages of reducing risk through minimizing exposure?

6. Through library research, identify databases and references of toxicological data and chemicals.

Chapter 3

1. Identify a biological feedstock and what product it is used to make.
2. Provide an example of a renewable and a depleting feedstock.
3. Are most chemical substances made from petroleum or renewable feedstocks?
4. Identify an example of how a depleting feedstock is collected. What environmental damage is associated with it?
5. What are some of the advantages of biological and renewable feedstocks?
6. What factors make a green chemical reagent environmentally beneficial?
7. What are the five aspects of a chemical process that should be part of a green chemistry evaluation?
8. What questions need to be addressed when selecting a solvent?
9. Why is real-time, in-process analysis beneficial to green chemistry?

Chapter 4

1. Define atom economy. Provide an example of a chemical reaction and calculate its atom economy.
2. Provide an example of a solvent that is currently used in industry and identify which, if any, environmental concerns there exists.
3. From a green chemistry standpoint, what would the advantages and disadvantages be of using a solvent that is volatile versus one that is not?
4. How is atom economy different from yield?
5. What are the most atom economical feedstocks and reagents?
6. What types of auxiliary substances are most commonly used in chemical manufacturing?

Chapter 5

1. Identify lists and databases for the toxicity of chemicals. [Hint: begin with government agencies such as Environmental Protec-

tion Agency, National Library of Medicine, Centers for Disease Control, and Occupational Safety and Health Administration.]
2. Identify an example of a direct and indirect toxic substance.
3. Which of the following are direct and which are indirect? Acute lethality, algael blooming, neurotoxicity, ozone depletion, endocrine disruptions, global warming.
4. Explain the difference between severity of a toxic effect and potency of a toxic chemical.
5. Give an example of a local environmental problem.
6. Give an example of a global environmental problem.

Chapter 6

1. What is a renewable feedstock?
2. Give an example of a renewable feedstock.
3. Give an example of a depleting feedstock.
4. How is recycling different from green chemistry?
5. What are the advantages and disadvantages of using carbon dioxide as a feedstock?

Chapter 7

1. Identify and analyze the atom economy of a well known reaction in chemistry [e.g. Friedel–Crafts, Wittig, Aldol].
2. Determine whether a waste is inherently generated in some well known reaction in chemistry [e.g. Friedel–Crafts. Wittig, Aldol].
3. Determine if an auxiliary substance is required for some well known reaction in chemistry [e.g. Friedel–Crafts, Witting, Aldol].
4. Rank the following reaction types in the order of most atom economical: Wittig, Dehydrohalogenation, Diels–Alder.

Chapter 8

1. Identify five methods of designing a safer chemical. Which of these require the greatest amount of toxicological data to utilize?
2. By accessing the Toxic Release Inventory data, identify how many pounds of toxic chemicals are released in your area per year, the largest releases, and the largest chemical by volume.

3. In the early part of the century, chlorofluorocarbons [CFCs] were considered safer substitutes in refrigeration. Identify the substances that CFCs replaced and the hazards it eliminated.
4. What is SAR?
5. Define bioavailability, how can one reduce bioavailability?
6. Define and give an example of mechanism of action.

Chapter 9

1. Identify a product made from synthetic chemistry that has had both dramatic advantages for the quality of life in the 20th century and has had negative environmental consequences.
2. Identify the recent Nobel Prize winning chemists who were recognized for their work on an environmental issue. Who were they? What was the environmental problem? What was their contribution?
3. Find additional examples of green chemistry in the chemical literature.
4. What are some of the alternatives to traditional solvents in chemical reactions?
5. Give an example of glucose used in the synthesis of a commodity chemical.
6. Give an example of the use of phosgene in a chemical synthesis. Give an example of green alternative to phosgene.

References

1. Breslow, R. (1997). *Chemistry Today and Tomorrow*. American Chemical Society, Washington, DC.
2. Office of Pollution Prevention and Toxics (1996). *1994 Toxics Release Inventory, Public Data Release*, Executive Summary, June, EPA-745-S-96-001, U.S. Environmental Protection Agency, Washington, DC.
3. Carlson, R. (1962). *Silent Spring*. Houghton Mifflin Co., New York.
4. Office of Pollution Prevention and Toxics (1996). *1994 Toxics Release Inventory, Public Data Release*, June, EPA-745-R-96-002, p. 34. U.S. Environmental Protection Agency, Washington, DC.
5. *Pollution Prevention Act of 1990* (1990). 42 U.S.C. §§ 13101.
6. Anastas, P. T. and Williamson, T. C. (1996). Green Chemistry: An Overview. In *Green Chemistry: Designing Chemistry for the Environment*. American Chemical Society Symposium Series, No. 626, (ed. P. T. Anastas and T. C. Williamson), pp. 1–17. American Chemical Society.
7. Anastas, P. T. and Farris, C. A. (1994). Benign by Design Chemistry. In *Benign By Design: Alternative Synthetic Design for Pollution Prevention*, American Chemical Society Symposium Series, No 577, (ed. P. T. Anastas and C. A. Farris), pp. 2–22. American Chemical Society.
8. Collins, T. J. (1997). Green Chemistry. In *Macmillan Encyclopedia of Chemistry*. Macmillan Inc., New York.
9. Trost, B. M. (1991). *Science*, 254, 1471–77.
10. The U.S. Department of Energy Office of Industrial Technologies (1993). *Alternative Feedstocks Program Technical and Economical*

Assessment. Thermal/Chemical and Bioprocessing Components (ed. J. J. Bozell and R. Landucci), p. 23.

11. The U.S. Department of Energy Office of Industrial Technologies (1993). *Alternative Feedstocks Program Technical and Economical Assessment. Thermal/Chemical and Bioprocessing Components* (ed. J. J. Bozell and R. Landucci), p. 23.
12. Varma, R. S., Chatterjee, A. K., and Varma, M. (1993). *Tetrahedron Lett.*, **34**, 46303–6.
13. Gross, R. A., Kim, J. H., Gorkovenko, A., Kaplan, D. L., Allen, A. L. and Ball, D. (1994). In *Preprints of Papers Presented at the 208th ACS National Meeting* Division of Environmental Chemistry, American Chemical Society, Washington, DC, **34**(2), pp. 228–9.
14. (a) Draths, K. M. and Frost, J. W. (1990). *J. Am. Chem. Soc.* **112**, pp. 1657–9. (b) Draths, K. M. and Frost, J. W. (1990) *J. Am. Chem. Soc.* **112**, pp. 9630–2. (c) Draths, K. M. and Frost, J. W. (1991) *J. Am. Chem. Soc.*, **113**, pp. 9361–3.
15. (a) Chang, V. S., Burr, B. and Holtzapple, M. T. (1997). *Appl. Biochem. Biotechnol.*, **63–5**, 3–19. (b) Office of Pollution Prevention and Toxics (1996) *The Presidential Green Chemistry Challenge Awards Program. Summary of 1996 Award Entries and Recipients*. July, EPA 744-K-96-001 p. 7. U.S. Environmental Protection Agency, Washington, DC.
16. (a) Stern, M.K., Hileman, F.D. and Bashkin, J.K. (1992). *J. Am. Chem. Soc.* **114**, pp. 9237–8. (b) Stern, M.K. and Cheng B.K. (1993). *J. Org. Chem. 58*, pp. 6883–8. (c) Stern, M.K., Cheng, B.K., Hileman, F.D., and Allman, J.M. (1994). *J. Org. Chem. 59*, pp. 5627–32.
17. Office of Pollution Prevention and Toxics (1996), *The Presidential Green Chemistry Challenge Awards Program, Summary of 1996 Award Entries and Recipients*. July, EPA 744-K-96-001 p.2. U.S. Environmental Protection Agency, Washington, DC.
18. (a) McGhee, W.D. and Riley, D.P. (1993). *Organometallics ll*, pp. 900–7. (b) McGhee, W.D., Riley, D.P., Christ, M.E. and Christ, K.M. (1993) *Organometallics* **12**, pp. 1429–33, and references cited therein. (c) Riley, D.P., Pan, Y., and Riley, D.P. (1994) *J. Chem. Soc. Chem. Commun.* pp. 699–700. (d) Waldman, T.E. and Riley, D.P. (1994). *J. Chem. Soc. Chem. Commun.*, pp. 957–8.

19. (a) Tundo, P., Trotta, F., and Moraglio, G. (1987). *J. Org. Chem.* **52**, p. 1300. (b) Tundo, P., Trotta, F., Moraglio, G., and Ligorati, F. (1998) *Ind. Eng. Chem. Res.* **27**, p. 1565. (c) Tundo, P., Trotta, F., and Moraglio, G. (1989) *J. Chem. Soc., Perkin Trans.* I p. 1070. (d) Tundo, P. and Selva, M. (1995) *Chemtech* pp. 31–35.
20. Komiya, K., Kukuoka, S., Aminaka, M., Hasegawa, K., Hachiya, H., Okamoto, H., Watanabe, T., Yoneda, H., Fukawa, I., and Dozone, T. (1996). New process for producing polycarbonate without phosgene and methylene chloride. In *Green chemistry: designing chemistry for the environment*, American Chemical Society Symposium Series, no. 626, (ed. P.T. Anastas and T.C. Williamson), pp. 1–17. American Chemical Society.
21. Collins, T. J. (1994). *Acc Chem. Res.* **27**, pp. 279–85.
22. Office of Pollution Prevention and Toxics (1996) *The Presidential Green Chemistry Challenge Awards Program. Summary of 1996 Award Entries and Recipients.* July, EPA 744-K-96-001 p. 30–1. U.S. Environmental Protection Agency, Washington, DC.
23. (a) Burk, M. J. (1991). *J. Am. Chem. Soc.*, **113**, p. 8518. (b) Burk, M. J. and J. E., F. Nugent, W. A., and Harlow, R. L. (1993). *J. Am. Chem. Soc.* **115**, p. 10125. (c) Burk, K. J. and Feng, S., Gross, M. F., and Tumas, W. J. (1995). *J. Am. Chem. Soc.* **117**, p. 4423. (e) Office of Pollution Prevention and Toxics (1996). *The Presidential Green Chemistry Challenge Awards Program. Summary of 1996 Award Entries and Recipients.* July, EPA 744-K-96-001 p. 38–9. U.S. Environmental Protection Agency, Washington, DC.
24. (a) Combes J.R., Guan, Z. and DeSimone, J.M. (1994). *Macromolecules*, **27**, pp. 865–7. (b) Guan, Z. and DeSimone, J.M. (1994), *Macromolecules, 27*, pp. 5527–32. (c) DeSimone, J.M., Maury, E.E., Menceloglu, Y.Z., McClain, J.B., Romack, T.J. and Combes, J.R. 91994). *Science*, **265**, pp. 356–9. (d) Office of Pollution Prevention and Toxics (1996). *The Presidential Green Chemistry Challenge Awards Program. Summary of 1996 Award Entires and Recipients.* July, EPA 744-K-96-001 p.11. U.S. Environmental Protection Agency, Washington, DC.
25. (a) tanko, J.M., Mas, R.H. and Suleman, N.K. (1990). *J. Am. Chem Soc.*, **112**, p. 5557. (b) Tanko, J.M., Suleman, N.K., and

Hulvey, G.A., Park, A., and Powers, J.E. (1993) *J. Am. Chem. Soc.*, **115**, pp. 4520–26. (c) Tanko, J.M., and Suleman, N.K. (1994). *J. Am. Chem. Soc.* **116**, pp. 5162–6. (d) Tanko, J.M. and Blackert, J.F. (1994). *Science*, **263**, pp. 203–5.

26. Office of Pollution Prevention and Toxics (1996) *The Presidential Green Chemistry Challenge Awards Program. Summary of 1996 Award Entries and Recipients.* July, EPA 744-K-96-001 p. 3. U.S. Environmental Protection Agency, Washington, DC.
27. (a) Breslow, R., Coinnors, R., and Zhu, Z. (1996). *Pure Appl. Chem.* **68**, 1527–33. (b) Breslow, R., and Zhu, Z. (1993). *J. Am. Chem. Soc.*, **117**, pp. 9923–4.
28. (a) Office of Pollution Prevention and Toxics (1996) *The Presidential Green Chemistry Challenge Awards Program. Summary of 1996 Award Entries and Recipients.* July, EPA 744-K-96-001 p. 8. U.S. Environmental Protection Agency, Washington, DC. (b) Paquette, L. In *Abstracts of Papers Presented at the 209th ACS National Meeting* (1995) Division of Organic Chemistry, American Chemical Society, Washington, DC, #123.
29. (a) Office of Pollution Prevention and Toxics (1996). *The Presidential Green Chemistry Challenge Awards Program. Summary of 1996 Award Entries and Recipients.* July, EPA 744-K-96-001 pp. 10–11. U.S. Environmental Protection Agency, Washington, DC.
30. (a) Office of Pollution Prevention and Toxics (1996) *The Presidential Green Chemistry Challenge Awards Program. Summary of 1996 Award Entries and Recipients.* July, EPA 744-K-96-001 p. 8. U.S. Environmental Protection Agency, Washington, DC. (b) Hurter, P.N. and Hatton, T.A. (1992). *Langmuir*, **8**, pp. 1291–9.
31. Epling, G.A. and Wang, Q. (1994). Preparative reactions using visible light: high yields from pseudoelectrical transformation. In *Benign by design: alternative synthetic design for pollution prevention*, American Chemical Society Symposium Series, No. 577, (ed. P. T. Anastas and C. A. Farris), pp. 2–22. American Chemical Society.
32. Kraus, G.A. and Kirihara, M. (1992). *J. Org. Chem.*, **57**, p. 3256.
33. (a) Grogan, J., DeVito, S.C., Pearlman, R.S., Korzekwa, K.R. (1992). Modeling Cyanide Release From Nitriles: prediction of

Cytochrome P450 Mediated Acute Nitrile Toxicity. *Chem. Res. in Tex.*, **5**, pp. 548–52. (b) DeVito, S.C. and Pearlman, R.S. (1992) *Med. Chem Res.*, **1**, pp. 461–5.

34. Office of Pollution Prevention and Toxics (1996) *The Presidential Green Chemistry Challenge Awards Program. Summary of 1996 Award Entries and Recipients.* July, EPA 744-L-96-001 p. 4. U.S. Environmental Protection Agency, Washington, DC.
35. Office of Pollution Prevention and Toxics (1996) *The Presidential Green Chemistry Challenge Awards Program. Summary of 1996 Award Entries and Recipients.* July, EPA 744-L-96-001 p. 30. U.S. Environmental Protection Agency, Washington, DC.
36. Office of Pollution Prevention and Toxics (1996) *The Presidential Green Chemistry Challenge Awards Program. Summary of 1996 Award Entries and Recipients.* July, EPA 744-L-96-001 p. 5. U.S. Environmental Protection Agency, Washington, DC.
37. Guarrera, D.J., Kingsley, E., Taylor, L.D., and Warner, J. C. (1997) *Proceedings of the IS&T's 50th Annual Conference. The Physics and Chemistry of Imaging Systems*, p. 537.
38. Hudlicky, T., Pitzer, K.K., Stabile, M.R., and Thorpe, A.J. (1996). *J. Org. Chem.*, **61**, 4151–53.
39. Guarrera, D., Taylor, L.D., and Warner, John, C. (1994). *Chemistry of Materials*, **6**, p. 1293.

Index

Made in the USA
San Bernardino, CA
20 December 2013